1005961980

Mixed Methods Research

MIXED METHODS RESEARCH

Exploring the Interactive Continuum

Carolyn S. Ridenour
Isadore Newman

Southern Illinois University Press
Carbondale

Copyright © 2008 by the Board of Trustees,
Southern Illinois University
All rights reserved
Printed in the United States of America

11 10 09 08   4 3 2 1

Library of Congress Cataloging-in-Publication Data
Benz, Carolyn R., 1942–
 Mixed methods research : exploring the interactive continuum / Carolyn S. Ridenour and Isadore Newman.
   p. cm.
 Includes bibliographical references and index.
 ISBN-13: 978-0-8093-2779-9 (pbk. : alk. paper)
 ISBN-10: 0-8093-2779-1 (pbk. : alk. paper)
 1. Research—Methodology. I. Newman, Isadore. II. Title.
Q180.55.M4B414 2008
001.4'2—dc22   2007031727

Printed on recycled paper. ♻
The paper used in this publication meets the minimum requirements of American National Standard for Information Sciences—Permanence of Paper for Printed Library Materials, ANSI Z39.48-1992. ∞

The authors gratefully acknowledge the help of Nicole Cleland, Samantha Posey, and Kaori Takano, graduate students at the University of Akron and the University of Dayton for their thoughtful input, feedback, and review in helping complete this book. Their gratitude extends to freelance copy editor Mary Lou Kowaleski, Barbara Martin, and project editor Wayne Larsen at Southern Illinois University Press for their expertise, thoughtfulness, and generosity in preparing this book for publication.

Dichotomies have their manifest utility, as well as their latent traps. They offer us an heuristic, an analytical scalpel, if you will, by which we can cut phenomena into slices thin enough for us to examine. This, of course, is a useful function, as long as we agree that analyzing the links, the many subtle membranes between the dichotomous end points, is a critical legitimate task for any serious analysis. In fact, if dichotomies are at all still useful in a modern world of concatenated complexities it is because the tension between the antithetically conceived end points represents the important possibilities for creativity, ambiguity, paradox, uncertainty, ambivalence, imagination, synthesis, and vision.

—Jean Lipman-Blumen, "The Creative Tension between Liberal Arts and Specialization"

Contents

List of Figures ix
Preface xi

1. Qualitative-Quantitative Research: A False Dichotomy 1
2. The Qualitative-Quantitative Research Continuum 16
3. Validity and Trustworthiness of Research 35
4. Strategies to Enhance Validity and Trustworthiness 67
5. Applying the Qualitative-Quantitative Interactive Continuum 91
6. "Science" and a Search for Principles of Practice in Mixed Methods 109

Appendix A: Phenomenological Research—Laughter and Humor 119
Appendix B: One Counselor's Intervention in the Aftermath of a Middle School Student's Suicide—A Case Study 123
Appendix C: Effect of Therapist's Self-Disclosure on Patients' Impressions of Empathy, Competence, and Trust in an Analogue of a Psychotherapeutic Interaction 130
Appendix D: The Monocultural Graduate in the Multicultural Environment—A Challenge for Teacher Educators 143
Appendix E: Teacher Reactions to Behavioral Consultation—An Analysis of Language and Involvement 160
Appendix F: An Example of Mixed Methods Criteria (Principles) for the Validity of Some Holistic Research Designs 176
Notes 179
Glossary 183
References 191
Index 203

Figures

1. The qualitative-quantitative continuum of educational research methodology conceptualized   22
2. Qualitative-quantitative mixed methods research as an interactive continuum   31
3. Links between research questions and truth value   38
4. Model of the consistency question to ask in critiquing research   93
5. Model of the consistency questions, the qualitative paradigm being dominant   95
6. Model of the consistency questions, the quantitative paradigm being dominant   96

PREFACE

Qualitative and quantitative research strategies and their underlying presuppositions have been increasingly debated since the early 1980s as though one or the other should eventually emerge as superior. We reject the dichotomy assumed by this debate. We address the growing popularity of the mixed methods research paradigm as one that most usefully assumes research as a qualitative-quantitative interactive continuum.

We take the position that the qualitative and quantitative philosophies are neither mutually exclusive (i.e., one need not totally commit to either one or the other) nor interchangeable (i.e., one cannot merge methodologies with no concern for underlying assumptions), which might be a consequence of mixed methods approaches that are not thoughtfully pursued. Rather, we present qualitative and quantitative research as interactive places on a methodological and philosophical continuum based on the philosophy of science as identified by Popper, Dewey, and Kerlinger and Lee. A researcher tests a theory, and, as results feed back to the original hypothesis, both inductive and deductive processes are operational at different points in time; qualitative and quantitative methods are invoked at different points in time; and feedback loops facilitate maximizing the strengths of both methodologies.

We intend this book to meet the general needs of two audiences: research designers and research consumers. For the first audience, it is imperative that researchers understand both the determinative nature of the research questions they ask and the assumptions on which they build their designs. This book is intended to assist in building effective designs. Although the first edition (Newman & Benz, 1998) focused on the *research question* as central to the research methods, in this second edition, we add the importance of the *research purpose* as even more important in each researcher's thinking and decision making. Both are crucial determinants. As for the second audience, sophisticated consumers of research need ways to assess the truth value of research findings. The practical approach to criticizing studies will enhance the quality of the judgments research consumers are able to make.

A unique contribution to research practitioners and consumers, the book addresses the growing role of mixed methods research in a new way—as an interactive continuum. The book is founded on the underlying philosophical assumptions of both qualitative and quantitative research. Both paradigms have their own contributions to building a knowledge base. The book serves mainly as a practical tool. Graphic depictions and narrative descriptions present research as a holistic endeavor; that is, both qualitative and quantitative paradigms coexist in a unified real world of inquiry.

Graduate students and social science faculty have already applied the ideas contained within this book, using drafts of our ideas over the past twenty years and in the first edition. The current volume would be most effectively used as a supplementary book in a graduate-level research-methods course. It could be used by faculty in the behavioral and social sciences to assist in their own research and their work with master's and doctoral students. Education and psychology are our areas of teaching and research, and the ideas are certainly applicable to these fields. However, the ideas and methods are also applicable to the fields of sociology, economics, political science, anthropology, business, and social work. Research workers outside the university will find this a useful supplement to other research manuals.

The purpose of this book is not to teach qualitative and quantitative methods. That is the purpose of other books. Our aim is to have consumers and planners of research think carefully about the consistency among research designs, research purposes, and research questions. We assume that the reader has had at least an introductory course in statistics. Fundamental conceptualization of research constructs will help the reader, but a comprehensive, in-depth understanding of neither statistics nor ethnographic strategies is necessary to use the ideas we propose. We have included a glossary to clarify those terms necessary to understand the interactive continuum concept.

Mixed Methods Research

Qualitative-Quantitative Research: A False Dichotomy

The research question initiates any research study. The research question is fundamental, much more fundamental than the paradigm (qualitative or quantitative) to which a researcher feels allegiance. In social and behavioral sciences, qualitative research is usually holistic, uncontrolled, exploratory, and carried out for purposes of understanding meaning. Quantitative research generally uses measured variables to test hypothesized relationships in more controlled situations. In the middle 1980s, the qualitative-quantitative dichotomy was being heavily debated, and discussion of the qualitative-quantitative debate began from that perspective—the primacy of the research question (Benz & Newman, 1986[1]). Subsequently, we built the model of the qualitative-quantitative interactive continuum. We persisted in holding onto the fundamental place of the research question as driving the researcher's decisions until after the first edition of this book was published in 1998. Then our perspective changed. A more scientific driving force, we concluded, is the research *purpose* (Newman, Ridenour, Newman, & DeMarco, 2003). Our threefold thesis in this book is that (1) the research purpose and the research question are the bases from which researchers make research design decisions, (2) validity is the framework through which one can assess the scientific quality of a research design, and (3) consistency among the research purpose, research question, and research methods establishes that validity.

This book describes our stance at a point in time, not final conclusions, which continue to emerge, to grow, and to build from our work as researchers and as teachers. The ideas in this book constitute a work in progress. Because the framework of the qualitative-quantitative interactive continuum presented here has been enlightening to colleagues and students for over twenty-five years, it might have value for contemporary research practitioners who work not only within the current context of frequently debated qualitative-quantitative research but also under pressure to consider mixed methods—a potential way to think about integrating both paradigms (or sets of methods) within a study.

Chapter 1 includes

- the history of qualitative, quantitative, and mixed methods research
- the typical purposes and outline of qualitative research studies
- the typical purposes and outline of quantitative research studies
- the emergence of mixed methods research
- the five qualities of science in educational research
- why the phrase "quantitative-qualitative research" is a false dichotomy

The Evolution of Three Paradigms

Qualitative and quantitative research methods have philosophical roots in the naturalistic and the positivistic philosophies, respectively. Qualitative researchers generally adopt an individual phenomenological perspective. On the other hand, most quantitative research approaches, regardless of their theoretical differences, tend to emphasize that there is a common reality on which people can agree. The debate between the two paradigms has been characterized as a "war" between very different ways of seeing and experiencing the world (see Tashakkori & Teddlie, 1998, for a summary of the paradigm wars).

For example, from a phenomenological and qualitative perspective, Van Manen (1990) and Geertz (1973) believe that multiple realities exist. Multiple interpretations from different individuals are equally valid. Reality is a social construct. If one functions from this perspective, how one conducts a study and what conclusions one draws from a study are considerably different from those of a researcher coming from a positivist position, which assumes a common objective reality across individuals. The extent to which commitments to these assumptions about reality are exclusive varies among qualitative and quantitative researchers. For instance, Blumer (1980), a phenomenological researcher who emphasizes subjectivity, does not deny that there is a stable reality one must attend to.

The debate between qualitative and quantitative researchers is based upon the differences in assumptions about reality, including whether or not it is measurable. The debate further rests on different beliefs about how we can best understand what we "know"—whether through objective or subjective methods.

The qualitative, naturalistic approach can be used when observing and interpreting reality with the aim of developing an explanation of what was experienced; an explanation might be considered a "theory." The quantitative approach is usually used when one begins with a theory (or hypothesis) and tests for confirmation or disconfirmation of that hypothesis.

It is important here to set the stage for not only abandoning the dichotomy but also to clarify how advocates of mixed methods have attempted to, in some way, integrate qualitative and quantitative research strategies. To begin, we examine a few of the key events in the evolution that established the qualitative-quantitative debate in the first place and how the potential of mixed methods has more recently come into that discourse. The debate may be but one more phase in the ebb and flow of an ever-changing philosophy of knowledge. To some, mixed methods may be a compromise, a way to integrate the qualitative and the quantitative paradigms. So also discussed in this chapter are the dangers of some applications of mixed methods as a potential panacea, a potential detour away from thoughtful, purposeful, and scientific research designs.

The genesis of the current qualitative-quantitative debate in educational research occurred as far back as 1844, when Auguste Comte claimed that the methods of natural science could be justified in studying social science (1844/1974; see also Vidich & Lyman, 1994). Science, in this view, is the collection and study of facts that can be observed through sensory input. These are the traditional data investigated by natural scientists, such as physicists, chemists, and biologists. According to this view, *true* science is accumulated through the study of phenomena that can be physically sensed, observed, and counted. The "unknowables," as Herbert Spencer described them in his 1910 essay, those things that cannot be sensed but might rely on reason or thought, are banished from scientific investigation. Both Comte and Spencer were positivists.

Interestingly, this "positivism" was a move away from a more speculative, more "unknowable" view, a move away from relying on theological and metaphysical explanations of the world. It was a move toward what could be "positively" determined (confirmed through sensory data). The philosophy maintained a grip on social science from the late 1800s through the early 1900s.

In the early 1900s, John Dewey, among others, questioned the absolutism of this position, viewing science as not separate and distinct from problem solving. His pragmatism considered science less rigidly than did the positivists. In *The Sources of a Science of Education* (1929), written some time after his initial speculations, he pointed out that practice should be the ground of inquiry. Learning, he claimed, was based largely on practice as the learner interacted with the surrounding world. He appreciated the deeper complexity of what educational and social scientists study. During the same period, a group of scholars who made up what became known as the Vienna Circle met and developed a new philosophy of science, logical positivism. Supporting Comte's positivism, they combined it with the symbolic logic of mathematics. Hypotheses derived using the rigor of mathematics (symbolic) could be combined with fact-gathering (positivism) to test their confirmability (which was eventually modified to *disconfirmability*). Although counter to Dewey's efforts to diffuse the positivistic assumptions, this hypothetico-deductive system was dominant in psychology and sociology in the middle years of the twentieth century. Education, which borrowed traditions of inquiry from these disciplines, was affected as well. The respect for precision in measurement and mathematical systems to test hypotheses and a quest for value-free science solidified this paradigm (Lagemann, 2000).

During the 1940s and 1950s, the quantitative paradigm dominated the social science and the educational research worlds. Behaviorists and organizational theorists utilized empirical fact gathering and hypothesis testing almost exclusively in studying educational and social phenomena. In the 1960s, a subtle shift away from positivism began due to the growing skepticism toward the domination of logical positivism and the evident chasm between human social systems and mathematical logic. New epistemologies began to emerge that acknowledged the value-laden nature of human social interactions. That human beings construct reality for themselves and that knowledge itself is transmitted in social ways were beginning to be asserted. Questions arose about the tenability of applying natural science methodology to complex human dynamics.

In 1962, in *The Structure of Scientific Revolutions*, the most significant work on this issue, Thomas S. Kuhn explored the shifts in science's dominant paradigms. His doctorate in theoretical physics led him to look back into the history of science as he sought to know more about its foundations. He describes how, by randomly exploring the literature, he was exposed to Jean Piaget and, in the late 1950s, to an historical analysis of social science and psychology. Kuhn's study of methodology drove him to leave physics and become a historian of science. He conceptualized the notion of paradigms, "universally recognized scientific achievements that for a time provide model problems and solutions to a community of practitioners," (1970, p. viii). He proposed that competing paradigms emerge chronologically when the dominant one no longer serves the explanatory needs of the scientific community. Using the context of physics from the perspectives of Isaac Newton and Albert Einstein, Kuhn explained these periods of competition, or scientific revolutions, in the natural sciences. He acknowledged that competing paradigms can possibly coexist on equal footing following such a revolution, or "paradigm shift," although, he cautions, it may be possible only rarely.[2] He proposed that the predominant paradigm affects researchers not only methodologically but also in how they see the world. Kuhn's conceptualization of "paradigm" has been reinterpreted by others, and many definitions are incorporated in the research literature.

Reaction to Kuhn was disparaging from both camps. The positivists feared he was undermining the dominant empirical world of science, and the postmodernists complained that he failed to destroy it. His controversial book ushered in an era of debate and dialogue about how researchers carry out their work and the assumptions of reality on which they rely. The debate between the empiricists and idealists[3] ultimately affected educational researchers as well.

The quantitative paradigm continued to reign over social science and prevailed in education until the mid-1980s. The strong traditional bias toward quantitative science seems consistent with Americans' preference for observable and countable facts, a sense that hard data are what science "is," a "Western" and technical way of thinking.

Logical positivism was losing supremacy in the 1980s. Concurrent with Kuhn's early notions of paradigms in the 1960s, society was undergoing radical changes. Some began to question the efficacy of the positivists' tools in explaining human organizational and social phenomena. Educators were acknowledging a more complex social context. Culbertson (1988), pointing to such 1960s' and 1970s' issues as racial integration, poverty, equal opportunity, the place of schools as tools in global economic competition, the Soviet Union's threat to our math and science preeminence, and the need to account for the success and failure of the nation's children, posits that, in this context of increased complexity, some began to search for policy tools beyond the quantitative paradigm. For many key decision makers, quantitative research had not been sufficiently successful in addressing important educational problems.

Recognizing that education served economic, political, and policy ends enhanced the opportunity for scholars interested in the culture of schools to begin to use anthropological strategies in their inquiry. These strategies fueled the interests of feminists, critical theorists, and others who sought to study schools as mediators of power and privilege. Policymakers' interest in the world of classroom practice grew. They increasingly expressed concerns that research and practice were unconnected and that this disconnection was in part due to the use of tightly controlled laboratory-like quantitative assumptions. Some social scientists began to derive theory from practice, rather than the other way around. For example, the 1954 Stanford Conference offered a first formal setting to explore how anthropological research strategies could be applied in schools (Lagemann, 2000).

Graduate programs preparing educational and social science researchers increasingly directed their attention, as did professional journals, toward qualitative research during the 1970s and 1980s. Allotting time and space to what had been considered the "alternative" paradigm led to wide discussions in the journals and at professional meetings. The editors of the *American Educational Research Journal,* for example, announced in 1987 that particular emphasis on qualitative methodology would be forthcoming as they evaluated manuscripts. The legitimacy of qualitative research was strengthened. A plethora of books, articles, and presentations on the trustworthiness of the qualitative paradigm materialized. Some extolled the virtues of qualitative research as the only avenue to "truth," while others claimed that only by holding onto the quantitative traditions can we have confidence in our knowledge base. The debate stimulated many questions: Which is more scientific: the deductive methods of the logical positivists (quantitative researchers) or the inductive methods of the naturalists (qualitative researchers)? Can the results of qualitative research be generalized as are the results of quantitative research? Can science be value laden (qualitative) or only legitimate if value free (quantitative)? What epistemological assumptions are violated by adopting one paradigm or the other?

Qualitative research methods are those generally subsumed under the headings ethnography, case studies, life history, narrative inquiry, field studies, grounded theory, document studies, naturalistic inquiry, observational studies, interview studies, and descriptive studies. Qualitative research designs in the social sciences stem from traditions in anthropology and sociology, in which the philosophy emphasizes the phenomenological basis of a study, the elaborate description of the "meaning" of phenomena from the perspectives of the people or culture under examination, verstehen. Often in a qualitative design, only one participant, one case, or one unit is the focus of investigation over an extended period of time.

Quantitative research, on the other hand, falls under the category of empirical studies, according to some, or statistical studies, according to others. These designs include the more traditionally dominant (in Western culture) ways in which psychology and behavioral science have carried out investigations. Quantitative modes have been the dominant methods of research in social science. *Quantitative* designs include experimental studies, quasi-experimental studies, pretest-posttest designs, and others (Campbell & Stanley, 1963; Shadish, Cook, & Campbell, 2002), in which control of variables, randomization, and valid and reliable measures are required and in which generalizability from the sample to the population is the aim. Data in quantitative studies are coded according to a priori operational and standardized definitions.[4]

Unlike many academic disciplines, educational research has never evolved into an academic community with common principles and canons of practice (Lagemann, 2000). Serious dialogue about the "science" and "research" of the field was delayed until forced upon researchers by political forces. Social science researchers have always represented diverse perspectives and multiple methods. This diversity of thinking comes from research questions that are generated by a diffuse profile of constituents across economic, political, social, academic, and legal communities. Logic suggests that diverse research questions about schooling require multiple methods of investigation. The questions of methodology raised in the qualitative and quantitative debate strengthened a multiple-paradigm approach in the 1990s.

According to Lagemann (2000), the need for both "decision-oriented" and "conclusion-oriented" studies was raised in 1969 by Cronbach and Suppes in a landmark meeting of educational thinkers. Their conclusion remains a need today.

> Decision-oriented studies are designed to help decision makers act intelligently; conclusion-oriented inquiries are designed to allow, through the free play of a researcher's imagination, for the discovery of new ideas, the description of previously hidden anomalies, and the investigation of relationships that had not been observed earlier. (Lagemann, 2000, p. 243)

The "war" has been a common metaphor used to characterize the qualitative and quantitative debate (Tashakkori & Teddlie, 1998). The *Educational Researcher*, a monthly publication of the American Educational Research Association (AERA), and the AERA annual meetings were the sites of ongoing debates in the profession (see, for example, Howe, 1985, 1988; Howe & Eisenhart, 1990; Miles & Huberman, 1984; Smith & Heshusius, 1985).

Since the mid-1990s, researchers have increasingly turned to mixed methods, combining qualitative and quantitative methods within a study. However, the discourse on mixed methods has rarely addressed qualitative and quantitative research as a continuum, the model since the 1980s. Tashakkori and Teddlie (1998) tell the story of the evolution of qualitative and quantitative research as the backdrop for mixed methods. Published in 1998, the same year as the first edition of this book, their focus on "pragmatism" ("what works") added substantively to the discourse in very different ways than did our model of a qualitative-quantitative interactive continuum. However, we agreed with Tashakkori and Teddlie, as they urged the dismantling the dichotomy of qualitative and quantitative paradigms.

The currency of qualitative perspectives, however, was politically weakened by federal legislation with the No Child Left Behind Act, 2001. NCLB triggered a debate into the meaning of scientific research in education and held up the randomized trial from medical research as the preferred model (the "gold" standard) for researchers seeking federal funding for education research. For almost a decade, the ways in which researchers can most appropriately study the dynamics of schooling have come to dominate the national discussion among education policymakers.

Even though mixed methods research has captured the attention of many educational researchers from the printing of our first edition to the current one, novice researchers continue to be prepared for "either-or" world, a dichotomous world of qualitative and quantitative research that might no longer exist. Too many students leave colleges and universities with a monolithic perspective. Either they become well-trained statisticians, or they become cultural anthropologists. If limited to only one or the other, they are equipped with only a narrow perspective and are methodologically weak in being able to ask and study research questions. Second, researchers in education and in the social sciences have not yet constructed a way to ensure protégés' success in utilizing both paradigms. Mixed methods research designs risk becoming the latest panacea if not scientifically applied (Ridenour & Newman, 2004; 2005). The interactive continuum model in this book builds the capacities of future researchers to incorporate a holistic conceptualization of research in their practice: qualitative, quantitative, and mixed methods research designs in ways that meet the criterion of being "scientific."

The *dichotomy* of qualitative and quantitative research is a false one. Although not an ontological construct, the dichotomy does serve a purpose. It allows separation of the ideas embraced within each paradigm. We slice the dichotomy thin to examine it and make the case in this chapter that the dichotomy does not exist in the scientific research realm.

Qualitative versus Quantitative: A False Dichotomy

All behavioral research is made up of a combination of qualitative and quantitative constructs. In this book, the notion of the qualitative-quantitative research continuum, as opposed to a dichotomy, is explored on scientific grounds. We believe that conceptualizing the dichotomy (using separate and distinct categories of *qualitative* and *quantitative* research) is not a productive way to think about research. The dichotomy is not consistent with a coherent philosophy of science. Rather than a dichotomy, it is a continuum and, as such, a coherent tool for making decisions about designing a study. A secondary theme is equally important: the interactive continuum is the best of the three models of mixed methods both for evaluating published research and for planning research. For example, what are known as qualitative methods can be beginning points, rich in-depth descriptions of a culture. This foundational strategy can be followed by quantitative methods to test hypothesized relationships within that culture. The sequence might be reversed. Hypothesized relationships about variables in the culture might be followed by rich in-depth descriptions in first-person accounts of those relationships.

A standard is needed to measure whether the qualitative, the quantitative, or a mixed methods continuum that includes both methodologies is the most appropriate process of designing a study to reach a level of truth. The standard of science gives an appropriate set of criteria.

Science: A Foundation for Research Design

The purpose of science is to explain natural phenomena. Science has many definitions but science, at its most basic level, is a way of knowing about the world, a way to get at "truth." On the other hand, there are various kinds of "truth," says Medawar (1984). This 1960 Nobel-prize–winning scientist in physiology and medicine writes of spiritual and religious truth as well as poetic truth (p. 4) and the fact of "scientific" truth—the result of the systematic processes of the scientist at work. He states that there is "no finally conclusive certainty beyond the reach of criticism. There is no substantive goal; there is a direction only, that which leads toward ultima Thule, the asymptote of the scientist's endeavors, the 'truth.'" (p. 5).[5] In other words, science has a heuristic purpose to generate knowledge. It is the heuristic value of research (and of science) that is seen as one of its most valuable contributions to behavioral research. Well-known paleontologist Mark Norell (2005) claims that there is no truth in science, "only the answers you have at the time," the self-correcting quality.

Other definitions render science a body of systematic knowledge. While there remain other ways of knowing about the world (e.g., literature, poetry, spirituality, and emotion), science is a highly respected way of knowing because the label *science* leads one to assume that the body of knowledge has been accumulated through, first of all, a systematic approach to collecting and analyzing the evidence. Not only is data collection systematic but also the reasoning of the researcher, and the planning by the researcher is systematic, organized, and logical. Krathwohl (2004) used the term *chain of reasoning* to capture the logic of the researcher. The term clearly connotes this systematic quality. *Systematic* implies that science is built through processes that are structured and sequential, planned and coherent (Rosenthal & Rosnow, 1991). The "science" in a specific field consists of an accumulation of knowledge in that field. The process of building that body of knowledge is the process of science, a process that conforms to systematic rather than haphazard procedures. Years ago, Lee S. Shulman (1987) characterized good educational research as "disciplined inquiry."

To be imbued with the label *scientific*, an endeavor must meet five criteria:

- It must be *systematic* in its processes and thinking. The study needs to be formal, systematic, organized, and prescribed.
- It must be *verifiable*. In other words, results of studies are testable. The truth value of scientific findings can be borne out by further testing by other researchers. Verifiability leads to the following characteristic.
- It must be *replicable*. Studies can potentially be replicated because of the basic systematic processes that science requires. Replicability is what a scientific body of knowledge accumulates through repeated tests of hypotheses or theories; the resultant knowledge is scientifically strengthened. With replication, findings can be confirmed and reconfirmed. Replicability imbues science with the next quality.
- It must be *self-correcting*. This implies that findings from replicated studies can overturn prior findings. Hypotheses may be discarded in favor of new hypotheses. According to Krathwohl (2004), "All scientific knowledge is held with a tinge of uncertainty—just enough that it could be replaced should more valid knowledge come to light. Knowledge that is replicated and reconfirmed is held with considerable certainty—enough that we act on it as though it were unquestionably true" (p. 51).
- It must *explain*. This characteristic, explanation of natural and human phenomena, is the traditional purpose of science, concern with examining variable relationships. Explanation, of course, is

the role played by theories—the requisite foundation of many scientific studies. Many scientific researchers not only target variable relationships but *causal* relationships, which embody the strongest aspirations of many researchers studying teaching and learning. It is these studies that are valued most highly by many researchers. For example, those seeking to raise student achievement and school success investigate the possible causes of such success.

We have purposely used the word *traditional* in this discussion so far. Science and all that the term *science* connotes have been almost exclusively linked to traditional positivist and quantitative research. So far, these descriptors are heavily weighted toward the deductive, objective, measurement-oriented world of the quantitative researcher. Qualitative research lies outside that realm, according to most of its adherents, at least insofar as it has not been aligned with science. But, we have been at a point of questioning that dichotomy (quantitative-qualitative, which parallels science and nonscience). We want to raise the question of how mixing qualitative and quantitative methods can fit within these scientific qualities.

Arguably, these five scientific qualities—systematic processes and thinking, potential verifiability, potential replicability, self-correction, and explanation—play a potential role in *all* educational research—perhaps even completely across the qualitative-quantitative continuum. Broadening how we think about research from a qualitative-quantitative dichotomy to a continuum that encompasses mixed methods raises this question: How do we accommodate the traditionally scientific and the traditionally nonscientific in ways that allow us to be coherent, consistent, and, indeed, completely scientific? Addressing this question encompasses the remainder of this book. Admittedly, the highly regarded status of science does not place science in a superior epistemological position. Ways of knowing other than science may be superior in some circumstances, depending on the need to know, the purpose of needing to know, and the context of the need to know (Bauer, 1992; Medawar, 1984).

The Nature of Both Science and Research in Education

Accommodating qualitative and quantitative research under a holistic umbrella of science might be achieved through not only a set of epistemological assumptions but also a set of procedural steps in designing a research study. Research and science are related endeavors. Research constitutes the process through which a scientific body of knowledge is accumulated. Research encompasses the activities of researchers as they carry out studies of phenomena in a particular field, for instance, in education. Research serves heuristic purposes in building a scientific knowledge base; new knowledge suggests possibilities for more questions and even newer knowledge. In education, the paradigms of quantitative, qualitative, and mixed methods research serve researchers' inquiry needs. Each of these paradigms needs to be briefly defined and put into context.[6]

Both positivism and naturalism, both empiricism and idealism, (i.e., both quantitative and qualitative research) are valuable to accumulating a knowledge base in education. Both contribute to the knowledge base. How could they not? Questions of interest about teaching and learning run the gamut from questions of cause and effect to questions of meaning. The science of education needs both perspectives to become a complete and coherent knowledge base, a scientific knowledge base.

Both quantitative and qualitative research must be able to fit within science in education if both methodologies serve to constitute the knowledge base in education. On the one hand, quantitative research in education rests on certain positivistic assumptions of reality—what traditionally has been categorized as the *scientific way of knowing about schools*. Knowledge about reality is assumed to be objective, separate and distinct from one who studies it; knowledge is deductively reasoned and generalizable; knowledge of reality is lawful, value free, and context free because reality is stable and knowable. Researchers approach the study of this reality through attempts to control settings and through theory testing, assuming a philosophy of empiricism. On the other hand, qualitative research rests on naturalistic and idealistic assumptions of reality—what traditionally has been categorized as the *nonscientific way of knowing about schools*. Knowledge about reality for qualitative researchers is built on an understanding of reality as holistic, dynamic, and irreducible to its particulars. Knowledge about reality is accrued subjectively, in natural settings that are value laden and context bound and that generate findings more difficult to generalize. Researchers approach the study of this reality through holistic means and a discovery orientation that builds theory rather than tests theory.

A counterexample to this theory is offered by Fontana and Frey (2005). They provide several types of interviews in the context of a qualitative research paradigm, posing that a structured interview to determine the facts of a situation (how many people oppose a nuclear-power facility in their neighborhood) can provide empirical data in the form of frequency counts that can be correlated with selected demographic variables. In this situation, "we can quantify and code the responses and can use mathematical models to explain our findings... we can speak in the formal language of scientific rigor and verifiability of findings" (p. 722). This example fails as an argument that qualitative research is *also* scientific; it succeeds in justifying that philosophical purposes and research situations dictate methods of data collection and analysis. The use of the interview in this situation is one better categorized, philosophically, as a quantitative study. We base our categories of qualitative and quantitative research on the bases of what purposes they serve rather than the nature of the data collected. Quantitative research is not necessarily defined by numerical data, and qualitative research is not necessarily defined by textual data.

Mixed methods research has offered a powerful new paradigm (Tashakkori & Teddlie, 2003). The danger is that some researchers might assume that after constructing domains of meaning from a qualitative study, they can code those themes as variables, test them empirically, and claim that they are using mixed methods. Unfortunately (or fortunately), it is not that simple for those procedures to rise to the level of science. The findings of qualitative studies (e.g., domains of meaning) and the findings of quantitative studies (e.g., probabilistic decisions about hypotheses) have different epistemological assumptions. Mixed methods are extremely valuable but cannot be a panacea (Ridenour & Newman, 2004).

This book contributes to the current discourse on qualitative, quantitative, and mixed methods research and assumptions underlying social science research by

- depicting an overall model of qualitative-quantitative interactive continuum that fits within one category of the currently accepted mixed methods paradigms
- suggesting ways to assess quality of published research

- providing a strong scientific context through principles based on consistency
- placing validity at the center of design decisions

Chapter 2 elaborates on the notion of the interactive continuum. In chapter 3, we discuss the central role of validity, review research methods, and address the strengths and weaknesses of quantitative, qualitative, and mixed methods research. Chapter 4 discusses ways of enhancing the validity of quantitative, qualitative, and mixed methods research, emphasizing qualitative research. Chapter 5 looks at four research studies—showing ways to analyze the consistency among the research questions, the methods, and the results. Chapter 6 contains a discussion of beginning principles of research practice, a preliminary set of tools that are a work in progress. These principles include questions to assess whether the research methods are consistent with research purposes and research questions.

All research in education stands on basic underlying epistemological assumptions. This is true for quantitative methods as well as qualitative methods. To the extent that these assumptions withstand the scrutiny of scientific inquiry, the methods can be supported, taught to novice researchers, and used professionally and ethically without reservation. Since the mid-1980s when quality in all educational professions came under public review, it has become particularly crucial to delineate the foundational bases of educational research. This book discusses such foundations.

THE QUALITATIVE-QUANTITATIVE RESEARCH CONTINUUM

Until the 1970s, any discussion of research methods presented them as dichotomized categories, either quantitative or qualitative. The two paradigms had been assumed to be polar opposites and, among some researchers, even separate and distinct scientific absolutes. Despite the strong historical roots of this dichotomy, an appreciation for mixed methods research has grown over the past two decades (Creswell, 2005; Frechtling & Sharp, 1997; Greene & Caracelli, 1997; Mertens, 2003; Reichardt & Rallis, 1994; Spicer, 2005; Tashakkori & Teddlie, 1998, 2003). Claims and counterclaims about the appropriateness of the two paradigms have been the genesis for a mixed methods approach to research. We assume research is conducted on a scientific foundation and that science is holistic. Because we assume it is holistic, we conceptualize science more broadly and in a less-compartmentalized way than those adhering solely to one or the other of the qualitative, quantitative, or mixed paradigms.

Knowledge about the world is gained in multiple ways. For example, an orderly, systematic investigation of objective reality may be combined with experiential and intuitive ways of knowing. Even though this spectrum is broad, science, to be called science, requires a fundamental set of systematic rules of procedure. Karl Popper (1962), in his earlier views, claimed that only those hypotheses that can lead to claims of falsifiability are scientific. That path, associated with quantitative research or the empiricists, may be too narrow. That science requires qualities of falsifiability as well as verifiability may not by themselves be sufficient. If they were, that view would exclude the metaphysical, the speculative, the existential, and the heuristic as legitimate ways of knowing. Diesing (1991) claimed that it would be better to admit all kinds of statements, both verifiable and falsifiable, into the realm of potential scientific investigation. We would go further and include the premise of the naturalists: the constructed reality that one interprets based on experience is included in what can be considered scientific. As a picture of lived reality, that knowledge, too, can be examined in scientific ways. Science is not only defined conceptually, it also embodies a set of rules of procedure.

This chapter presents a conceptualization of research methods as existing on an interactive continuum rather than as a qualitative-quantitative dichotomy. Included are discussions of scientific inquiry, the purpose of research, the kinds of questions that are typically posed, and our fundamental assumption that each research question is derived from a purpose and that the research question and purpose together dictate the research method. We argue that thinking through qualitative and quantitative assumptions is always involved to at least some degree in every research study (Tashakkori & Teddlie, 2003). We embrace a notion of mixed methods that resists dichotomizing qualitative and quantitative research and accepts them, rather, as places on an interactive continuum, situated as they relate to "theory."

Chapter 2 includes description of

- the central roles of the research purpose and the research question in designing a research study
- the place of "theory" in qualitative and quantitative research
- the link between postpositivism and the qualitative-quantitative interactive continuum
- three categories of mixed methods research
- two conceptual models of the qualitative-quantitative interactive continuum: one that explains the philosophy and one that explains the sequence of methodological decisions

Science as a Set of Systematic Procedures

Science consists of systematic and organized processes (as opposed to random or haphazard processes), and it allows acquisition of knowledge toward truth in a variety of ways. We assume no singular epistemology. We do assume a singular *process* to think through the research design decisions. No one method to acquire knowledge is superior. With these rules (and their underlying assumptions serving as standards), one can define ways of making decisions about research. One can have confidence in the findings that result. Researchers must ultimately determine whether the qualitative, the quantitative, neither, both, or a continuum including both methodologies is most effective in fulfilling the purpose of the study and addressing the research question. A systematic approach to addressing research problems is necessary no matter which ideology or epistemology one holds.

First, the researcher must begin with the nature of the research question in concert with the research purpose. Both may be considered iteratively; that is, the research purpose may generate the research question; the research question may lead to refinement of research purpose. The research question must be addressed in the context of the purpose of the study (Newman, Ridenour, Newman, and DeMarco, 2003). Why the study is being conducted, the purpose, must be clearly understood so that the research design and the methods will serve the intended needs of the researcher and his/her audiences. Without a clear purpose or set of purposes, implications of the results will be difficult to render.

Secondly, identifying the evidence needed to address the question needs to be identified as well as the underlying epistemological assumptions of that needed evidence. In other words, to address the research question and to fulfill the research purpose, what epistemological stance must be taken: a particularistic or holistic stance? An inductive or deductive stance? An objective or subjective stance? What epistemological assumptions am I, the researcher, adopting in this research study?

Third, decisions about research design and the nature of evidence follow. Results of these deliberations will lead to determining whether the evidence is or is not quantified, according to the design of the study. In other words, the decision about what evidence to collect as well as what to do with that evidence after it is collected should be dictated by the research question and purpose. Fourth, decisions about the source of evidence, the setting, the timing, the measures or lack of measures, and analysis of evidence are made. Fifth, plans for communicating results to audiences in order to fulfill the research purposes are made.

This systematic set of steps is discussed later in this book in more detail, but here they are presented to show that considering science to be holistic and heuristic does not permit researchers to proceed haphazardly. The qualitative-quantitative dichotomy no longer exists. The decisions about methods are based on a holistic spectrum of possibilities, are inclusive, and follow naturally from the research question and purpose. For example, Miller and Lieberman (1988) characterize a "new synthesis" in education. In their review of studies of school improvement, they acknowledge the different sets of assumptions underlying qualitative and quantitative studies but describe studies that combine the technological perspective of the quantitative with the cultural perspective of the qualitative.

The paradigm of positivism (quantitative research) continues to dominate social and behavioral science. It is steeped in historical tradition. For one thing, the training of research methodologists in social science and education has been heavily weighted on the side of quantitative research designs and statistics. The challenge from qualitative adherents over the past thirty years has not been successful in overthrowing that dominance but has led to the debates between the advocates of quantitative research and the advocates of qualitative research.

Instead of an us-them dichotomy, however, the scientific tradition can be strengthened when science is both positivistic and naturalistic in its assumptions. Two fundamental epistemological requirements are made of the researcher: one must clearly and openly acknowledge one's assumptions about what counts as knowledge and maintain consistency in the links between those assumptions and the methods derived from them. We argue that this consistency makes the research scientific. As clarified in chapter 5, it is the consistency among the research question, purpose, and methods that ensures a study is scientific, not the choice of one paradigm or the other.

Qualitative Research Conceptualized

In the third edition of their qualitative research handbook, Denzin and Lincoln (2005) acknowledge that the term *qualitative research* means different things to different people. They offer what they call a "generic definition."

> Qualitative research is a situated activity that locates the observer in the world. It consists of a set of interpretive, material practices that make the world visible. These practices transform the world. They turn the world into a series of representations, including field notes, interviews, conversations, photographs, recordings, and memos to the self. At this level, qualitative research involves

an interpretive, naturalistic approach to the world. This means that qualitative researchers study things in their natural settings, attempting to make sense of, or to interpret, phenomena in terms of the meanings people bring to them. (p. 3)

Qualitative data have been defined by Patton (1990) as "detailed descriptions of situations, events, people, interactions, observed behaviors, direct quotations from people about their experiences, attitudes, beliefs, and thoughts and excerpts or entire passages from documents, correspondence, records, and case histories" (p. 22). Denzin and Lincoln (2005) go much further: qualitative research is not restricted to a set of methods or the nature of the evidence:

> Nor does qualitative research have a distinct set of methods or practices... Qualitative researchers use semiotics, narrative, content, discourse, archival and phonemic analysis, even statistics, tables, graphs, and numbers. . . . [and also use] ethnomethodology, phenomenology, hermeneutics, feminism, rhizomatics, deconstructionism, ethnography, interviewing, psychoanalysis, cultural studies, survey research, and participant observation, among others. (p. 7)

Unlike some researchers who characterize qualitative research as evidenced by words and quantitative research as evidenced by numbers, Denzin and Lincoln differentiate the two paradigms based on assumptions of reality. That foundation also serves as the basis for the model of mixed methods that we promote here: the qualitative-quantitative interactive continuum.

In contemporary research literature, writers have ascribed many meanings to the word *theory* that take it beyond its traditional scientific meaning. We contrast two meanings of the word here to clarify how we are using the concept in the mixed methods interactive continuum. We use the word *theory* in this discussion in the sense that Popper (1959) conceptualized it. A theory is a scientific explanation of phenomena that is made up of testable hypotheses. Only hypotheses that are falsifiable fall into the category of scientific theories. Science requires that a hypothesis be constructed in such a way that if it is false, it can be eliminated. Hypotheses that are stated in ways that preclude being eliminated if they do not hold up to contradictory data are unscientific hypotheses. Figure 1 shows that quantitative researchers frequently begin with theory, a scientific (falsifiable) explanation of relationships among variables that the researcher wishes to test. In that same figure, we suggest that some qualitative researchers might construct an explanation (theory) in their findings, a set of hypotheses that are scientific. For example, Glaser and Strauss (1967) base their qualitative approaches on this notion of theory—falsifiability, explanation, and prediction. This relationship between qualitative and quantitative research is at the core: quantitative research as theory testing and qualitative research as theory building.

However, the word *theory* has a different meaning in some qualitative-research literature. Theory is used to refer to a perspective or world view that includes the personal assumptions and insights one has about the world and his/her place in it. Because the qualitative researcher is often cast as the "instrument" of data collection and analysis (e.g., Patton, 1990, p. 56), that researcher's assumptions and insights about the world and about the data impinge on those data processes and are "part of the data" (Patton, 1990, p. 58). Examples of such perspectives (sometimes referred to as "theoretical" frameworks or "paradigms") might include critical theory, feminism, Marxism, and queer theory (Denzin & Lincoln, 2005, p. 24). The use of the word *theory* to mean a personal perspective on the world is not the way we use the word *theory* in the interactive continuum.

A qualitative researcher using grounded theory builds theory from the data. Theory is therefore grounded in the data rather than being abstract or tentative, according to the pioneers, Glaser & Strauss (1967). Compared to a researcher who fulfills his/her purpose by testing hypotheses, a grounded theorist approach avoids issues of data collection and is applied as a data-analysis technique. Instead of coming from the conceptual level to the empirical level, one would begin at the empirical level (data analysis) and end at the conceptual level (theory construction). According to Charmaz (2000), more recent attacks on grounded theory come from critics' claims that its methods are consistent with positivism and empiricism. For example, two frequent criticisms are the assumption that data are "objective" and that reality can be captured and recorded (Charmaz, 2000). A proponent of this method as a strong qualitative strategy, however, Charmaz maintains that grounded theory offers "a set of flexible strategies, not rigid prescriptions" (p. 513).

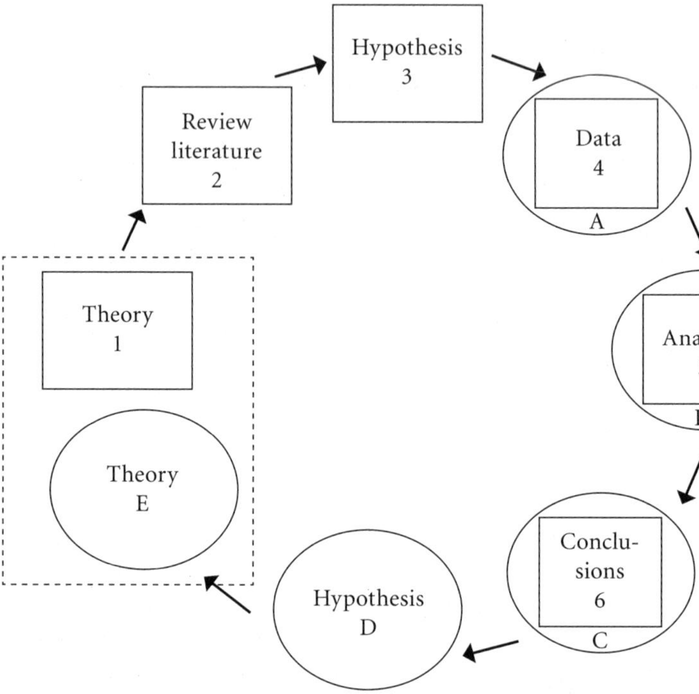

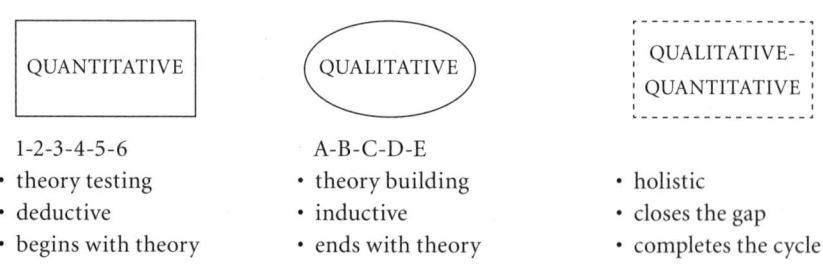

Conceptually, in *this* model, the theory is neither at the beginning nor at the end—but the square (the quantitative) and the circle (the qualitative) overlap and continue the cycle, closing the qualitative-quantitative gap. Neither the squares nor the circles make a whole. (See definition of "theory" on page 20.)

QUANTITATIVE	QUALITATIVE	QUALITATIVE-QUANTITATIVE
1-2-3-4-5-6	A-B-C-D-E	
• theory testing	• theory building	• holistic
• deductive	• inductive	• closes the gap
• begins with theory	• ends with theory	• completes the cycle

Figure 1. The qualitative-quantitative continuum of educational research methodology conceptualized

Inductive reasoning and deductive reasoning are both subsumed under scientific inquiry, yet they characterize a distinction between purely qualitative and purely quantitative methods. Patton (1990) states the separation even more strongly: "The cardinal principle of qualitative analysis is that causal relationships and theoretical statements be clearly emergent from and grounded in the phenomena studied. The theory emerges from the data; it is not imposed on the data" (p. 278).

Quantitative Research Conceptualized

Quantitative research is frequently referred to as *hypothesis-testing research* (Johnson & Christensen, 2004; Krathwohl, 2004). Investigating the effects of a treatment or an intervention is typical of this paradigm. For example, using deductive logic, studies begin with statements of theory from which research hypotheses are derived. Then an experimental design is established in which the variable in question (the dependent variable) is measured while controlling for the effects of selected independent variables. Randomly selecting participants for the study is desirable to reduce error and to cancel bias. The sample of participants is selected to represent a defined population. After the pretest measures are made, the treatment conducted, and posttest measures made, a statistical analysis reveals findings about the treatment's effects, that is, whether or not the results are likely due to sampling error alone. To support repeatability of the findings, one experiment usually is conducted, and statistical techniques are used to determine the probability of the same differences occurring over and over again. These tests of statistical significance result in findings that confirm or disconfirm the original hypothesis. Theory revision or enhancement follows. This would be a true experiment. These procedures are deductive in nature, contributing to the scientific knowledge base by theory testing. This is the nature of quantitative methodology. Because true experimental designs require tightly controlled conditions, the richness and depth of meaning for participants are usually sacrificed. As a validity concern, this may be a limitation of quantitative design.

Replication is the key to science; a single study generally cannot add to the knowledge base. Newman, McNeil, and Fraas (2004) assert that attention to the issue of statistical significance has been overblown in the literature; the more important concern is that the data are replicable.

To enhance the scientific quality, researchers should include an estimate of replicability in their research reports.

Mixed Methods Research Conceptualized

Mixed methods is the third paradigm. Again, our purpose here is to raise questions about how education research is now and can continue to be scientific. We also are attempting to argue that validity or trustworthiness is the lens through which standards of practice for mixed methods research might be developed. In this section, we conclude that mixed methods research designs might ultimately be built on postpositivist assumptions.

Our qualitative-quantitative interactive continuum and the procedural steps in this model are closest to Denzin and Lincoln's notion of postpositivism (2000), which they describe, among other things, as an attempt to accommodate a classical Campbell and Stanley (1963) approach[1] within both quantitative and qualitative research (p. 14). What Denzin and Lincoln refer to as a "modified dualist" understanding of qualitative and quantitative research we can accept as at the core of what is a holistic paradigm, a continuum that allows multiple methods (or single methods) to be selected based on the purposes of each research study.

Denzin (1994) describes the four responses that have been made to the legitimation crisis, the crisis that questioned how qualitative research can be evaluated. First, the positivists apply the same four criteria to qualitative research as to quantitative research: "internal validity, external validity, reliability and objectivity" (p. 297). Second, the postpositivists believe a separate set of criteria needs to be developed for qualitative research. Denzin characterizes those who fall in this group as often creating a set of criteria that parallels that of the positivists but is adjusted to naturalistic research. Third, the postmodernists claim that there can be no criteria for judging qualitative research. Fourth, according to Denzin, the critical poststructuralists believe that new criteria, *completely different from* those of *both* the positivists and the postpositivists, need to be developed. It is with this last group, the critical poststructuralists, that Denzin aligned himself in the 1994 volume.

Within Denzin's structure, our position aligns with his second category, postpositivism, because we believe a different set of criteria should be applied to assess qualitative research. The criteria established by Lincoln and Guba (1985) differ from those established for quantitative research, but they are philosophically derived from them. Denzin (1994) describes the legitimation crisis as the concern for the validity of qualitative research, with the postpositivists calling for a set of rules of procedures to establish validity.

> A text's authority, for the postpositivist, is established through recourse to a set of rules that refer to a reality outside the texts. These rules reference knowledge, its production and representation. . . . Without validity (authority) there is no truth, and without truth there can be no trust in a text's claim to validity (legitimation). (p. 29)

Postpositivism was the impetus for mixed methods research, according to Tashakkori and Teddlie (1998). Postpositivists blur the lines separating positivist and naturalist philosophies. Postpositivism replaced positivism—an epistemology that failed to withstand a barrage of skepticism for a variety of reasons (Phillips & Barbules, 2000), for example, the inconsistency across many of its underlying assumptions. Positivists assume there is an ultimate knowable reality, but that is inconsistent with the assumption that researchers can know only the reality that is observed and counted (Phillips & Barbules, 2000).

Postpositivists concluded that a "disinterested scientist" was also untenable. Postpositivists, according to Tashakkori and Teddlie (1998), believe that the personal values of researchers influence the object of their study. They also contend that facts are always value laden and that one constructs meaning from the reality of one's own experience. Tashakkori and Teddlie (1998) claim that these tenets of postpositivism are "shared by" both qualitative and quantitative researchers (p. 8), a conceptual break with what was previously understood, that is, that these are tenets of the naturalist, typically associated with qualitative research alone.

Postpositivism is possibly less stable and bounded than other epistemologies, according to Phillips and Barbules (2000).

> The new approach of postpositivism was born in an intellectual climate . . . an "orientation," not unified "school of thought," for

there are many issues on which postpositivists disagree. But they are united in believing that human knowledge is not based on unchallengeable, rock-solid foundation—it is *conjectural* [italics in the original]. We have grounds, or warrants, for asserting the beliefs, or conjectures we hold as scientists, often very good grounds, but these grounds are not indubitable . . . warrants for accepting these things can be withdrawn in the light of further investigation. (p. 25–26)

This last phrase—"warrants . . . can be withdrawn in light of further investigation"—seems to be consistent with the self-correcting nature of science. Replication and confirmation build a scientific knowledge base. Because postpositivism is "nonfoundational," that is, knowledge has no solid foundations, the researcher works with the best warrants he/she has at the time. One could then argue that qualitative and quantitative research paradigms are compatible. This compatibility has encouraged the paradigm of pragmatism, according to Howe (1988), a point of view many have adopted.

Categories of Mixed Methods

All studies that apply to both qualitative and quantitative methods are not necessarily alike; they are not all in the same category. Mixed method studies are categorized into the nonintegrative, the simultaneous attempt, and the interactive continuum. From reviewing scores of studies and studying the designs of many mixed methods advocates, we constructed these categories for explanatory purposes; this nomenclature helps explain the value—specifically, the *scientific* value—of mixed methods. These categories are helpful in examining the possibility of principles of using mixed methods. They do not form a continuum, but we have structured them along a conceptual dimension, that is, a quasi-continuum of mixed methods designs ranging from those weaker conceptually to those stronger conceptually. One might construe the continuum scientifically as well. In other words, those mixed methods designs that are conceptually weaker are also scientifically weaker. Concomitantly, those designs that are conceptually stronger are also stronger in approaching the scientific benchmarks (systematic, verifiable, potentially replicable, self-correcting, and intended to explain phenomena).

Nonintegrative

In the nonintegrative type, qualitative research is carried out, followed by the use of quantitative methods, or the other way around, without having either method informing the other. The two methods are used independently without integrating them or linking them to common purposes. This may be the type most frequently cited in the literature.

Simultaneous Attempt

In the simultaneous attempt type, the researcher attempts to carry out qualitative and quantitative methods simultaneously and both with the same purpose(s). This generates virtually insurmountable epistemological problems because underlying assumptions of qualitative and quantitative research studies are very different. For instance, quantitative research assumes some type of objective reality from which one can generalize from a sample to a population with some estimate of confidence in doing so; whereas, in qualitative methods, no objective reality is assumed. Reality is unique for each individual. One cannot generalize from a sample to a population with any estimate of confidence, nor should one be interested in doing so. Qualitative and quantitative assumptions are incongruent with one another. Conflicting assumptions cannot be held at the same time in interpreting the same data for a common purpose.

Interactive Continuum

The interactive continuum is the third category, based on the qualitative-quantitative interactive continuum. This is the one with which we are most comfortable (Newman & Benz, 1998). This category of mixed methods is different from the first two in a number of ways. The first two categories dichotomize quantitative and qualitative methods, while this third conceptualization rejects the dichotomy and relies on a continuum in which research may be predominantly qualitative or predominantly quantitative. We prefer to characterize this type of mixed methods as holistic because this term diminishes the notion of a dichotomy. In this third type, the methods are driven by the research questions linked to the purpose(s). Identifying the research question and the research purpose (or, because there may be more than just one, the questions and purposes) is accomplished through an iterative process. The researcher moves from the question to the purpose and back through both iteratively to exhaust all possible questions and possible purposes. Once all potential purposes are identified and all potential research questions linked to those purposes are articulated, the researcher designs the strategies for collecting and analyzing evidence. That evidence must be defended as consistent with the purposes and questions. That evidence may rest epistemologically within the qualitative and/or the quantitative paradigm. The focus of the researcher is predominantly on the research purpose and the research question (Newman et al., 2003). The design (whether qualitative, quantitative, or mixed methods) is a consequence of the more important focus on purpose and question. The paradigm decision is a logical conclusion, not a starting place. The choice of paradigm or paradigms is a result of thoughtfully reflecting on (and, ultimately, clarifying) the purpose and the question. This sequence strengthens the conceptual clarity of the research study.

When novice education researchers learn first and foremost to focus on their research purposes and their research questions (and *not* only on the methods), they are much more likely to avoid conceptual confusion in their research.

When considering methods from both ends of the continuum (qualitative and quantitative) and their scientific base (their basis in what we call *knowing repeatable facts),* different assumptions are apparent. The concept of a continuum is a more comprehensive schema. Evidence of such a continuum is demonstrated by an increasing number of researchers who apply multiple methods to their research and by the increased popularity of multimethod approaches in sociological research[2] (Tashakkori & Teddlie, 2003). Despite the debate, these ideas are not new, but they are now more strongly emphasized. More than thirty years ago, Mouly (1970) alluded to multiple-perspective research as

> the essence of the modern scientific method. . . . Although, in practice, the process involves a back-and-forth motion from induction to deduction, in its simplest form, it consists of working inductively from experience to hypotheses, which are elaborated deductively from implications on the basis of which they can be tested. (p. 31)

If we accept the premise that scientific knowledge is based upon verification, the contributions of findings derived from a qualitative (inductive) or quantitative (deductive) perspective can be assessed. It then becomes clear how each approach adds to the body of knowledge by building on the findings derived from the other approach. This is the premise of the interactive continuum. A schema that depicts the philosophies of this continuum appears in figure 1.

The place of theory in both philosophies is shown to overlap. This is where the concept of the continuum is most clear. For the qualitative researcher, the motivating purpose is often *theory building;* while for the quantitative researcher, the intent is often *theory testing.* Neither the qualitative research approach nor the quantitative research approach encompasses the whole of research. Both are needed to conceptualize research holistically.

The schematic in figure 1 cannot symbolize all qualitative and quantitative studies, but what it can symbolize is a conceptualization or way of thinking about some kinds of qualitative and quantitative studies. In general, the qualitative researcher follows the sequence shown in the circles and labeled with letters A through E). At circle A, data are collected, interpreted, absorbed, and experienced. At circle B, the data are analyzed; and at circle C, conclusions are drawn. From those conclusions, a hypothesis is created (circle D). This hypothesis can be used to develop theory (circle E), the goal of the qualitative research question in some instances.

Quantitative research begins with theory (square 1). From theory, prior research is reviewed (square 2); and from the theoretical frameworks, a hypothesis is generated (square 3). This hypothesis leads to data collection and the strategy needed to test it (square 4). The data are analyzed according to the hypotheses (square 5), and conclusions are drawn (square 6). These conclusions confirm or conflict with the theory (square 1), thereby completing the cycle.

The qualitative-quantitative continuum is strengthened scientifically by its self-correcting feedback loops. In every research study, the continuum can be symbolically conceptualized as an organizing tool, a chain of reasoning for researchers to make links between and among their research purposes, questions, and methods.

When one conceptualizes research in this way and uses the built-in feedback mechanism, positive things happen that are less likely to occur in a strictly qualitative or a strictly quantitative study. For example, data may be more parsimonious in a quantitative study if the research question has emerged from a participant observation, historical review, or series of interviews. These qualitative foundations of a quantitative study enhance its validity. These empirical materials may, for example, become forces driving the data-collection instruments or identifying the sample to be selected.

Although probably no single representation or schematic diagram can easily explain the concept of the qualitative-quantitative interactive continuum, figure 2 presents the model conceptually and summarizes the interrelationships between qualitative and quantitative methods as approaches to scientific inquiry. It is important that the reader understand that this is a simplification of a concept that has an infinite number of combinations. As shown on figure 2 all research endeavors probably start out with a purpose and a topic of interest. Researchers are obligated to justify why this topic and this purpose is of value. Studies need to be justified as serving one or more than one purpose. A typology of research purposes can serve as a tool for researchers to identify the purpose or purposes of their investigations (Newman et al., 2003). For example, some researchers are interested in testing the impact of an innovative treatment; other researchers are interested in exploring some unknown phenomenon; still others are interested in delving deeply into the causes of some historical event. Sometimes, this speculation becomes formally structured and takes on the qualities of a theory. However, it can remain loose and informal, based on phenomenological experiences and assumptions. Generally, once the speculation stage is reached, the next step, in both qualitative and quantitative research, is to do a review of the literature. However, there are certain qualitative researchers who believe that one should not enter the research with preconceived notions, that the data should be free from the bias of the researcher's prior knowledge and expectations. Consequently, a literature review is not desirable. Two examples from the literature demonstrate this view.

Frederick Erickson (as quoted in Goetz & LeCompte, 1984) describes one group of advocates for ethnographic studies who enter the field purposefully assuming a naïveté, while others merely suspend their preconceptions. L. M. Smith (1967) describes how one assumes ignorance in terms of the foreshadowed problem. Like Erickson, some qualitative researchers believe the study can begin without prior knowledge; one deliberately avoids learning anything about the topic or setting. However, it is obvious that this is impossible to achieve due

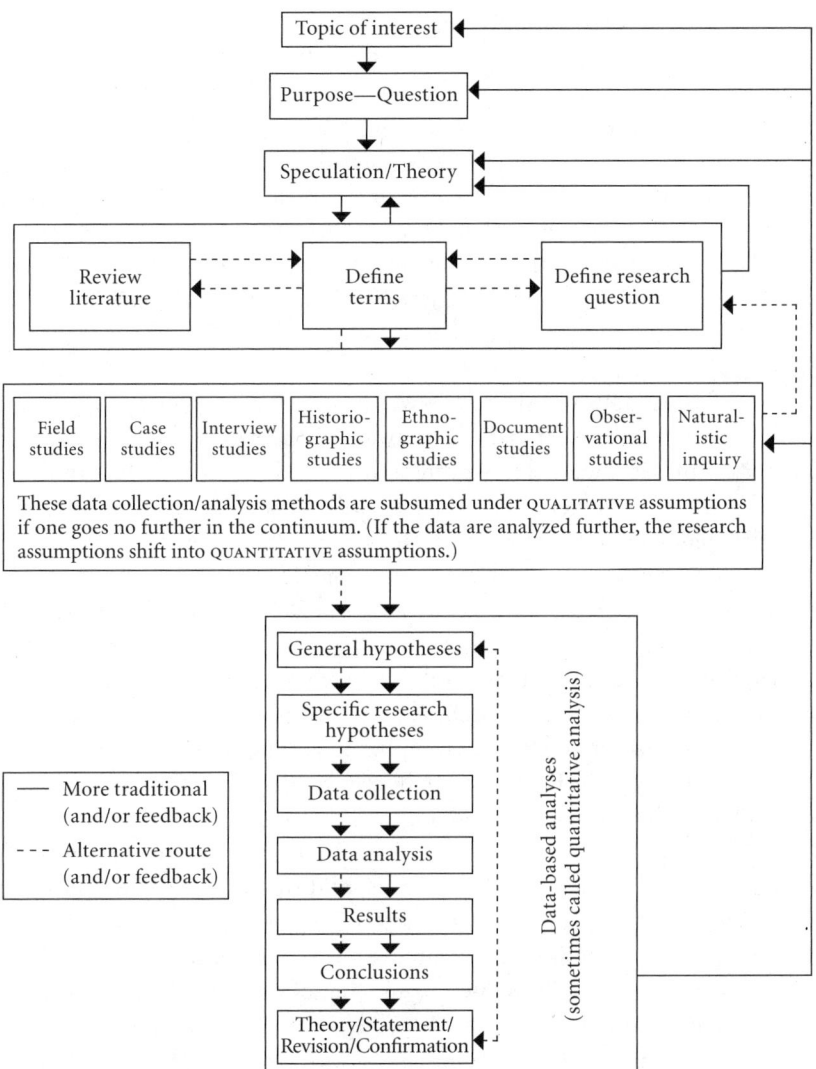

Figure 2. Qualitative-quantitative mixed methods research as an interactive continuum

to the research purpose one must establish. Therefore, Smith might claim the problem can be one that is foreshadowed at least; in other words, a working hypothesis about what might be "out there" in the field drives the research question. This problem keeps the researcher on the track of the most cogent data. While one is in the field, the research question guides what one attends to; this strategy has become common for qualitative researchers. We see this concept of foreshadowing as not entirely different from the notion among empiricists of *working hypotheses*, defined as those relational statements derived from descriptive research, theory, or personal experiences (Rosenthal & Rosnow, 1991; see also Ary, Jacobs, & Razavieh, 1990).

We argue, however, that one always has preexpectations and that it is important for researchers to be aware of their own biases. If aware of these biases, the researcher might more likely control for bias in the data-collection stage. This is the rationale for the schematic structure presented here. At the same time, the reader must understand that this diagram is an attempt to conceptually represent the qualitative *and* the quantitative strategies within systematic scientific inquiry. The decision of method rests on the research question's purpose and assumptions, which guide the research method—*not* vice versa. The method should not dictate whether the research is qualitative or quantitative; the reader should not interpret figure 2 as implying that it does.

The review of the literature can be related directly to the topic, to the historical background or chronology of events and studies surrounding the topic, or to the applications and usefulness of the topic. Often the literature review, definitions of terms, and the research question are interdependent. One is an outgrowth of the others or, depending on how much information the researcher has at the beginning, one tends to change the others. This interdependence of these three elements and speculation and theory is represented by dotted lines in figure 2.

The next box in figure 2 depicts the qualitative methods. It is difficult to represent these methods accurately as discrete entities because overlap almost always occurs. One study strategy (e.g., case study) may use another study strategy (e.g., focus groups) within its framework as well as within its data-collection procedures. For example, if an investigator uses an ethnographic strategy, the collected information might be coded numerically and analyzed statistically in a hypothesis. However, an underlying assumption of the ethnographic method is that one cannot generalize; the researcher cannot begin with a purpose toward generalizability of findings and then carry out the research methods in ways that disallow generalizability.

In quantitative research, a researcher seems to directly proceed from reviewing and defining to developing hypotheses and collecting data. This is represented in figure 2 by the dotted line descending from the "review-literature–define-terms–define-research-question" box that bypasses the qualitative-strategies box into the quantitative-methods box. In quantitative analysis, this bypassing is called the *derivation of hypotheses*. These derivations may be more appropriately considered qualitative analyses in simplified form. The researcher examines the literature and, based upon this process, derives theoretical expectations, which become the derived hypotheses. The solid line going from the "review literature–define terms–define research question" box to the qualitative-strategies box and its feedback loop is what some individuals will identify as qualitative analysis in its entirety. Other researchers would suggest that one go from that feedback to the quantitative methods box and use it before appropriate and scientific conclusions could or should be made from qualitative data. As one can see, the qualitative analysis with its feedback loops can easily modify the types of research questions that will be asked in quantitative analysis; and the quantitative-analysis results and its feedback can change what will be asked qualitatively. Therefore, this model is not only a continuum from qualitative to quantitative but interactive.

In a paper given at American Educational Research Association twenty years ago, we presented an example of the need to study the world holistically, an example that is relevant today (Benz and Newman, 1986). Over several semesters, an on-campus late-afternoon seminar received consistently low mean ratings from student teachers. It was not until telephone interviews were conducted that it was revealed that the content of the seminar was highly valued, and the professors' feedback was sorely needed, but the time the seminar met was most disturbing to the students because the time conflicted with some of their school responsibilities. Numerical ratings alone masked the real value of the seminar, but by adding the interviews, a more holistic understanding was possible. Student teachers' quantitative ratings of their experiences on questionnaires (quantitative research) were followed by interpretive analyses of their personal experiences (qualitative research).

One needs to identify qualitative and/or quantitative research according to the purpose of the study and the question being asked. If one wishes to terminate the discourse in the scientific process within the qualitative-analysis box of this schema, then the research is qualitative. One goes no further in the diagram. If one utilizes the strategies in the quantitative-analysis sequence, the research is quantitative.

In the diagram, one can see the feedback loops that facilitate theory revision, see where theory fits in both methods, and, to some extent, understand why theory is never proven absolutely. It is always subject to modifications as new data enter the system. This approach fits and is applicable to conceptualization in both qualitative and quantitative research. Examples of research critiques presented in chapter 5 demonstrate how one study might productively lead to other investigations.

In the last twenty-five years or so, proponents of both approaches have assumed that one or the other paradigm would eventually "win."[3] Advocates of mixed methods approaches vary from those who maintain that both sets of epistemological assumptions can be held simultaneously to those who argue that research methods can be entirely divorced from concerns for epistemology.[4] The real issue is improving the quality of research. The focus of the rest of this book is the application of the continuum model in concrete ways to help researchers conduct their own and evaluate their own and others' research.

VALIDITY AND TRUSTWORTHINESS OF RESEARCH

Validity is the truth value of a research study and, therefore, a central concern for all researchers. In their most recent qualitative research handbook, Denzin and Lincoln (2005) define important contemporary research issues, and, unsurprisingly, they claim, "An important topic may be one that is widely debated (or hotly contested)—validity is one such issue" (p. 192). This chapter is devoted to validity.

Those who read and rely on research outcomes must be satisfied that the studies are valid, that they have led to truthful outcomes. Threaded throughout this book is one strong theme: research methods must be driven by the research questions and purposes. Within this mandate lies the essence of validity—the focus of this chapter. Winter (2000) captures this essence when he characterizes the important link between "purpose" and "method" as the crux of validity: "Many of the allegations of invalidity from both [qualitative and quantitative] sides can be attributed to a failure to recognise the purposes to which each methodology is suited" (p. 8).

This chapter begins with the argument that connecting research purposes, questions, and methods is central to validity and truth value. Next discussed are validity issues in quantitative research through a review of internal- and external-validity criteria and, lastly, issues of validity in qualitative research, a much more amorphous topic. Contrasted with the classic criteria for the validity of quantitative research described by Campbell and Stanley (1963) that most researchers accept, the criteria of validity (legitimation) of qualitative research has no consensus. Selected ideas here come from Hammersley (1992), Kvale (1995), Lather (1995), LeCompte and Goetz (1982), and Tierney (1993) to help locate our own views as being closest to Lincoln and Guba's (1985) postpositivist perspective. We list Lincoln and Guba's (1985) criteria for enhancing truth value. These are the most useful criteria for designing and critiquing qualitative studies. The chapter concludes with a discussion of the construct validity in mixed methods studies.

This chapter includes discussion of

- validity and trustworthiness
- design validity and measurement validity
- threats to internal validity and to external validity
- ex post facto research
- conceptualizations of validity in qualitative research
- qualities or design components that add to the trustworthiness of qualitative research
- validity issues raised by using mixed methods

The Meaning of Validity

The notion of *validity* has a strong consensus among most traditional education researchers. The concept is applied in at least two contexts—in research design (internal and external validity) and in measurement (the validity of the measurement). We struggled to conceptualize validity in these two contexts in our model. We ultimately placed the validity of measurement within the category of Campbell and Stanley's "instrumentation" (1963, p. 5). When there is a strong estimate of content validity and construct validity of one's measurement, for example, the threats to the internal validity of one's design are lessened.

Wolcott has conceptualized the same two validity concepts, but he does so in a chronological way. In his essay in Eisner and Peshkin's 1990 edited volume, *Qualitative Inquiry in Education,* Wolcott claims to be unconvinced that validity needs quite so much emphasis in naturalistic studies (1990). He suggests that the word *understanding* replace *validity* in qualitative research. Interestingly, he chronicles the evolution of the emphasis on validity over the past three decades as a developmental pattern: "test validity" predated the notion of "validity of test data," then followed, in sequence, a focus on "validity of test and measurement data," "validity of the research data on tests and measurements," the "validity of research data," and, finally, the "validity of research" itself.

Our concern has been validity of the research itself: the truth value of research outcomes; and we deliberately used the words *validity* and *trustworthiness* in the title of this chapter. Validity has traditionally meant an estimate of the extent to which the data measure (or the design measures) what is intended to be measured. But not all researchers conceptualize the links between design and truth value in this way. *Validity,* as we have defined it in our experience, does not fit all researchers' definitions. *Trustworthiness* is a recent term that we borrow here to relate to a broader notion of truth value. In essence, it can be considered somewhat parallel with our notion of validity.

Without validity, according to Denzin (1994), there is no truth, and without truth, there is no claim of validity. He has described a recent phenomenon in qualitative research as a "crisis of legitimation" (p. 295), in which the traditional standards of validity, reliability, and objectivity do not apply. Legitimation implies that the research methods are consistent with the philosophical underpinnings of the question. For example, the positivist assumes an objective reality; the postmodernist assumes no objective reality and no objective truth. To a certain extent, the notion of legitimation mirrors our set of consistency questions.

Although a perfectly accurate portrayal of our notions of validity across the continuum is not possible, we can outline the major dimensions of our model. As much as a diagram is able, figure 3 depicts our reasoning about connecting purposes, questions, methods, and truth value.

Connecting Research Questions, Methods, and Truth Value

Figure 3 should not be interpreted to mean that the decisions embedded within it are necessarily made in this linear fashion. On the other hand, the research question and the research purpose *always* initiate any set of decisions the researcher makes. The purpose of the study and the question driving the study are iteratively considered; they jointly determine the decisions about methods. The decisions may be cyclical. (Quantitative and qualitative methods may cross over in particular situations, thus the dotted line connecting them in figure 3). What is not negotiable is the structural connection among the question, purpose, methods, and truth value. While design decisions are "emergent" in good qualitative research, the researcher's thinking is not emergent. The conceptual linking of purpose-question-methods-truth value must be defensible and predetermined (Ridenour, Newman, DeMarco, and Newman, 2003).

As mentioned before, the purpose and the research question guide what methods a researcher selects. For that reason, figure 3 begins with the research question and research purposes boxes, which leads to the

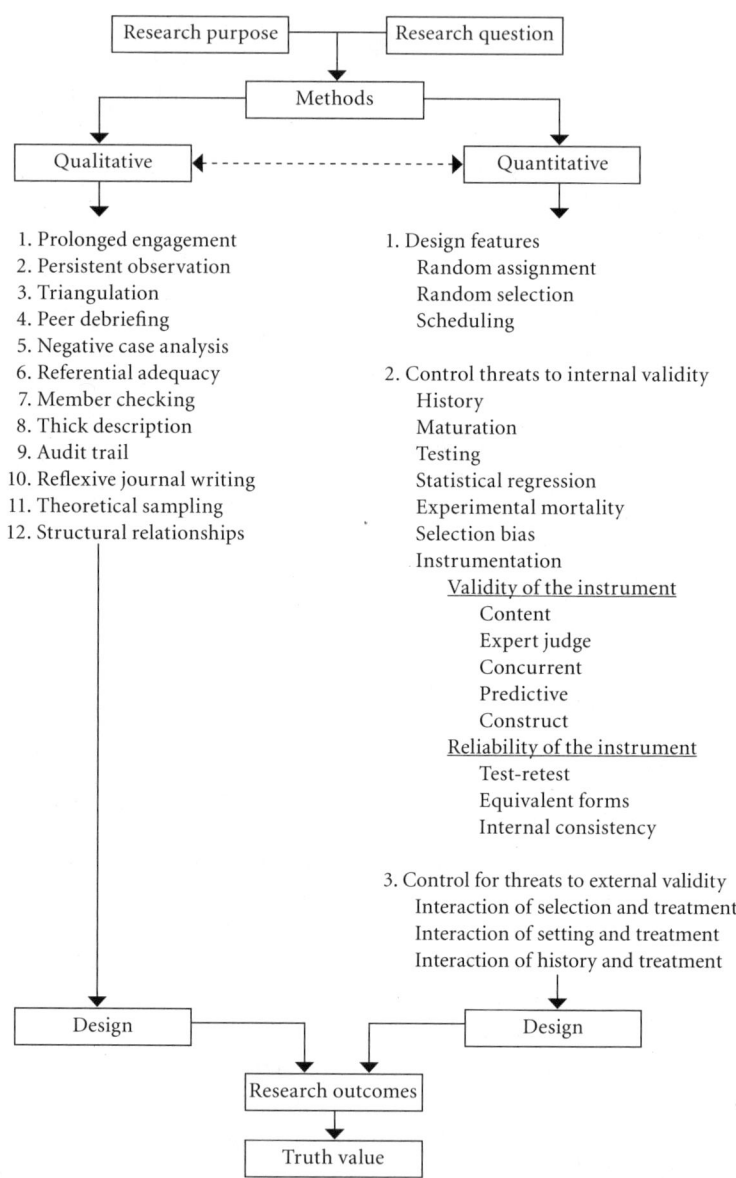

Figure 3. Links between research questions and truth value

next construct—methods. Because our overall intent is to present research holistically, not as a dichotomy, we wanted to discuss both paradigms in similar conceptual ways as much as possible. Methods are those features that make up the content of a researcher's decisions, thus the label. The methods box leads into the two paradigms, the qualitative and quantitative boxes. In some instances, a researcher may adopt mixed methods—utilizing both qualitative and quantitative methods. Only the mixed methods model in which the two paradigms are sequential is acceptable because it is the only model of the three that links research question and purpose and methods in a way that fulfills scientific assumptions of consistency.

We adopt the traditional issues for validity in quantitative research. These concerns about validity include both external and internal validity, on the one hand, and measurement validity, on the other. Both are generated by the need to have confidence that test, data, or design do measure or reflect or produce what we intend it to measure, reflect, or produce.

Validity concerns in qualitative research position us closest to what Denzin (1994) describes as the *postpositivist perspective;* that is, there needs to be a separate set of criteria for assessing qualitative research. The postpositivists call for a set of rules or procedures to establish validity. These criteria, from Lincoln and Guba (1985), might serve as a set of benchmarks, similar to the criteria developed by Campbell and Stanley (1963) for monitoring threats to design validity in quantitative research. Figure 3 presents some of those design strategies that methodologically support the legitimation of qualitative research.

Hammersley (1992) claims one type of truth (validity) is reliability. This argument is founded on different philosophies rather than on different methodologies. Our position is that validity cannot exist without reliability. By definition, validity estimates the extent to which the test or set of data or design actually measures or reflects or produces what it is supposed to measure, reflect, or produce. The basic purpose of reliability is to help researchers estimate validity. Reliability is an estimate of measurement error (Newman, Newman, Brown, & McNeely, 2006). Therefore, if one has validity, there is no need for estimates of reliability.

Polkinghorne (1991) asks whether the findings of the study are believable. Then he defines validity as the correspondence between findings and "reality." Others, such as Lincoln and Guba (1985), prefer the term *credibility* to the term *validity*.

Our emphasis in exploring the validity of research methods across the qualitative-quantitative continuum is in enhancing the truth value of the outcomes of the research we conduct. Figure 3 shows the reasoning that connects methods to the truth value of those outcomes. Each methods list leads to a point labeled "design." All the methods together create the design.

Enacting the design in each paradigm leads to the next point on the figure, "research outcomes," which are common to both paradigms. From the outcomes, finally, comes "truth value," the last point on the figure. The *truth value* label, while originating with qualitative research methodology, is synonymous with the validity of the research results. To some extent, it is what Polkinghorne is asking: Are the findings of the study believable or true?

Measurement Validity and Design Validity in Quantitative Research

There is an important difference between measurement validity and design validity. The first, measurement validity, estimates how well the instrument measures what it purports to measure. The second, design validity, encompasses internal and external validity. Measurement validity falls under Campbell and Stanley's (1963) "instrumentation." Internal validity is the extent to which any causal difference in the dependent variable can be attributed to the independent variable. External validity is the extent to which the results of the research study can be generalized to other settings or groups.

Measurement and design validity are not independent. A research design is only internally valid if it has measurement validity and reliability. This also is true for the ability to generalize: estimates of the stability of the measurements are needed to estimate external validity. Measurement validity is a subset of internal validity. As discussed in the next section, strong measurement validity diminishes the threats to internal validity that come under Campbell and Stanley's category of "instrumentation." We chose to separate measurement validity and design validity to clarify the discussion.[1] Next, we discuss internal and external validity as they relate to quantitative research; then we discuss validity in qualitative research.

Design Validity: Internal and External Validity

Design validity for quantitative research traditionally has been addressed through the concepts of internal and external validity. A similar concern should be addressed for those studies that are predominantly on the qualitative end of the continuum. Within the context of a holistic view, however, we subsume all of these consistency questions into the concept of *methods* validity, as shown in figure 3.

Internal validity is defined as the extent to which the researcher is able to claim the independent variable causes the effects of the dependent variable (Newman et al., 2006). LeCompte and Goetz (1982) ask this question to get at internal validity: "Do scientific researchers actually observe or measure what they think they are observing or measuring?" (p. 43). To the extent that they are observing and measuring what they think they are, they have validity. The second conceptual area is external validity, defined as the extent to which the results of a study apply to other people, groups, times, and places (Newman et al., 2006). LeCompte and Goetz (1982) ask this question to assess external validity: "To what extent are the abstract constructs and postulates generated, refined, or tested by scientific researchers applicable across groups?" (p. 43).

Other attempts to achieve such internal validity include counterbalancing across the treatment conditions (Newman et al., 2006). This design involves taking two learning tasks, for instance, and giving group 1 the tasks in *A-B* order and group 2 the tasks in *B-A* order. In this case, the order is used as an independent variable, and its influence on the variability in the dependent variable can be isolated and measured. Measuring the independent variable assists the researcher in obtaining internal validity because its effects can be separated from the treatment effects.

Without internal validity, one can only reason that the approach used to answer the question of interest is capable of estimating the relationship, and no statement about causation is possible. Even though there are those among the ranks of qualitative researchers who say they are not interested in internal validity, those who wish to infer causal relationships must be concerned with it. Even some who dismiss this concern as being only a quantitative researcher's dilemma will admit to processes like triangulation and theoretical sampling, which can be conceptual attempts to get at internal validity.

External validity reflects the extent to which the design and the data match the world. While tight controls over the variables of interest increase internal validity, they tend to do so at the expense of external validity—that is, the more laboratory-like the conditions the more precise and valid are the measures. However, the world is not tightly controlled. Variables operate in the world outside the laboratory. Sampling of participants across several strata that reflect the world to which the results will be generalized (e.g., age, socioeconomic status, occupation) increases external validity. The inability to control independent variables in ex post facto research (while it tends to decrease internal validity) tends to increase external validity because such variables are more relevant to their real-world distribution (Newman et al., 2006).

Diagrams used to illustrate research designs help the researcher plan, interpret, and analyze. Symbols, as defined here, are used to depict the components of the research strategy and are applied in chapter 5. While the following are examples of symbols commonly used, they do not include all possibilities, nor are they universally used.

- X   treatment or experimental treatment; refers to the experimenter's treatment for a group
- –X   no treatment or absence of treatment (one group might receive treatment, a second group not; second group is commonly called the *control group*)
- Ⓧ   independent variable, not manipulated and either attribute or assigned variable
- O   measurement or observation, frequently some type of test score; can have any number of subscripts
- R   random assignment of participants to groups
- M   assignment of participants to groups using matching
- M_r assignment of participants to groups by first matching participants and then randomly assigning each matched participant to group

Threats to Internal Validity

According to both the classic work of Campbell and Stanley (1963) and the more recent Shadish, Cook, and Campbell (2002) update of that classic, internal validity is limited when an assortment of factors is uncontrolled.

History. History is a factor that could account for a change in a group. *History* means any nontreatment, extraneous event that intervenes between the pretest and posttest measurements. For example, a school district investigates the effect of a new set of mathematics textbooks on achievement by measuring achievement at the beginning and at the end of the year. During that year, however, all the students move to a new building. This is an example of a history factor that might affect the results of the study. The researcher, in this case, would not know if the new books or the new school caused any difference in student achievement outcomes measured in the study.

Maturation. Another factor that could account for a change in a group is *maturation,* which means any growth or development that would normally take place independent of an experimental treatment. An example would be an analysis of the impact of a school lunch program on students' growth. Continued growth normally would occur as part of their maturation. In this case, the researcher's task is to identify what change is due to the lunch program and what change is due to ongoing maturation. History and maturation are often confused because each is related to something occurring in the time between pretest and posttest. Historical effects are external to the participants, while maturation effects are internal. Other examples of internal changes would be psychological, such as boredom, or physiological, such as fatigue.

Testing. Factors associated with measuring devices, *testing* factors, can also cause change to occur. This can happen when a pretest sensitizes people to an experimental treatment and actually causes them to behave differently during that treatment. Suppose a teacher is evaluating the impact of a character program on the moral development of students. If the teacher gives a pretest that asks questions about morality, the pretest could get the students concerned about the topic and thus influence them, making them more receptive to the program than they might have been without the pretest. This same phenomenon can take place with other factors, such as achievement. Testing effects are the results that the first test has in sensitizing participants to the treatment, which then affects the posttest.

Instrumentation, including concerns for measurement validity. On the other hand, the term *instrumentation* refers to the effects that are due to unreliable measurement instruments. If one has an unreliable pretest measurement, any change noted in the posttest measurement might be due to the instability of the measurement device rather than to the treatment. Instrumentation is a threat to internal validity and is related to the validity and reliability of the data. The data the researcher uses as evidence are gathered through various instruments that measure the variables of interest.

Two concerns of the researcher when collecting data by means of a measuring instrument are the validity and reliability of the instrument (Linn & Miller, 2005). Weaknesses in either threaten internal and external validity in this instrumentation category. This section discusses measurement validity first and follows with a discussion of reliability.

A test or measurement instrument has *face validity* to the extent that it appears to the individuals being assessed to be measuring what it purports to be measuring. This is generally considered a poor estimate of validity. It might be important because if a test does not have face validity (credibility), it can decrease the likelihood of people participating or volunteering.

When experts in the content areas make subjective judgments about the validity of the instrument, there is said to be *expert-judge content validity*, which has also been called *logical validity* and, times, *definitional validity*. It is similar to face validity, but it generally is estimated by using a table of specifications. The specifications include content items the test is supposed to measure and the measures of how well and how completely the items represent content areas.

How well one assessment instrument correlates with an already established valid assessment instrument is known as *concurrent validity*. *Known-group validity* is a type of concurrent validity. It is estimated by how well the instrument differentiates between two known groups. If the instrument is supposed to measure successful marriage, the instrument should be able to distinguish between two groups of people who have been identified previously as successfully or not successfully married. To the extent the instrument does this, it has known-group validity.

An estimate of how well an instrument predicts a future assessment outcome is called *predictive validity*. The major difference between concurrent validity and predictive validity is that concurrent validity is demonstrated contemporaneously with one's research measurement while predictive validity is future oriented. Sometimes concurrent and predictive validity are combined and are then called *statistical, empirical,* or *criterion validity.*

The most important and the most difficult to estimate is *construct validity.* It is used to estimate is how well the instrument measures the underlying construct it attempts to measure, and it is generally estimated by the use of a combination of the other types of validity mentioned. *Factor analysis* is a statistical technique used to estimate construct validity,

A claim of measurement validity is never absolute or complete but is always an estimate. Claiming a measure is "valid" is a judgment about what implications or predictions can be made from the data; it is not a scientific absolute. Because measurement validity is, by definition, limited, so, too, is truth value always limited.

If validity is confirmed, having reliability is implicit; however, it is possible to have reliability without validity. The major purpose of reliability is either to support or to improve validity. Reliability describes consistency, while validity estimates how well a study or a set of instruments measures what it purports to measure. Reliability estimates tell whether the outcomes will remain stable over time (i.e., whether they are "repeatable") or whether they are consistent among independent observers (i.e., whether different observers will report the same outcome).

In quantitative research, reliability in data collection is assured in three ways: measuring internal consistency, applying test-retest correlation coefficients, and using equivalent forms of the instrument. If reliability is not assured, then the scientific assumption of accuracy of measurement is violated, that is, the facts are not repeatable.

Just as reliability is estimated by calculating the internal consistency of a test form, a similar measure can be derived from a *structured-interview schedule.* Control over the timing, the environment, and the question order is possible where no such control is possible with some questionnaires. To the extent that these controls enhance validity, they fulfill reliability requirements by definition. For nonstructured interviews, no such reliability estimates are possible. One cannot have validity without reliability, and, concomitantly, to the extent that one has validity, one need not estimate reliability. For this, among other reasons, the predominant focus in this book is validity.

Statistical regression. Whenever groups of participants are selected on the basis of extreme scores and for no other reason, a phenomenon called *statistical regression* occurs. If participants are selected for a study solely because they score extremely low, they will tend at posttest to score higher, regardless of the treatment. The opposite is also true—the higher scorers will posttest lower. Both are examples of extreme cases regressing toward the mean of the population. Thus, significant differences between pretest and posttest scores can occur because an extreme was initially selected.

Mortality. The loss of research participants between a pretest and a posttest is called *experimental mortality*. If a sample of one hundred participants scores an average of 95 on an IQ pretest, and, for some reason, fifty lower scoring participants leave the study before posttesting, the average IQ of the remaining fifty people could be 150. Therefore, the differences between pretest and posttest scores could be, likely, not due to the treatment but due to the differential loss of participants between pretest and posttest (i.e., experimental mortality).

Selection bias. When participants are assigned to two or more comparison groups, and not all groups are given the treatment, there is *selection bias*. If these groups are different before treatment, then any difference between pretest and posttest scores may be due to the initial differences rather than to the treatment. An example is assigning individualized instruction (treatment) to highly motivated children and assigning traditional instruction to children with little motivation. If the groups were tested for gains at the end of a unit, any difference found might be due to initial motivation differences rather than to treatment differences.

Ex Post Facto Research

The two types of independent variables are *active* and *attribute* (Kerlinger & Lee, 2000; Newman et al., 2006). Active variables are under the control of the researcher and can therefore be manipulated. Attribute variables, such as gender and race, cannot be manipulated. If all the independent variables are nonmanipulatable, then the research is defined as *ex post facto*. Ex post facto research is sometimes relegated to an inferior position among types of research design. The terms *ex post facto research* and *correlational research* are sometimes used interchangeably. For example, Gall, Gall, and Borg (2005) use the label "causal comparative" design for those situations in which researchers are prevented from manipulating the independent variable, but group comparisons are desired. They distinguish causal-comparative designs from correlational designs based on the situation that in the former the investigator studies only two variables at a time, and in the latter, three or more variables might be simultaneously analyzed.

In correlational research, causation cannot be inferred. Many methodologists warn against possible misinterpretations of research in which the experimenter does not have control over the independent variables. Some consider ex post facto research to be exploratory (Newman et al., 2006).

In ex post facto research, causation is sometimes improperly inferred by those who assume that one variable is likely the cause of other because it precedes it or because one variable tends to be highly correlated with another (e.g., smoking—the independent variable is assumed to cause cancer—the dependent variable). Although a correlated and preceding relationship is necessary, it is not sufficient for inferring a causal relationship. To conclude that a causal relationship exists, all other explanations for the effect on the criterion (dependent variable) must be controlled for, and the only possible explanation for changes in the dependent variable must be due to the independent variable under investigation—a situation of internal validity.

Only with a true experimental design does one have the experimental control to achieve internal validity (Kerlinger & Lee, 2000; Newman et al., 2006). Ex post facto research lacks this control for a variety of reasons. One cannot randomly assign and manipulate the independent variable because it has already occurred and is not under the control of the researcher. Another common weakness is that the design is not capable of controlling the confounding effects of self-selection. For example, suppose one conducts research to see what effect early childhood training has on motivation. Also suppose that a significant relationship is found between early independence training and later adult motivation. One might, therefore, incorrectly conclude that the independence training causes this adult motivation. Another explanation might be that volunteer participants who have had independence training are more likely to demonstrate a greater degree of adult motivation. What causes this motivation might be more related to what causes the participants to volunteer for the study than to the independence training.

If the research purpose is to show cause-effect relationship, clearly ex post facto research is inappropriate. However, if the question deals with relationships, it is appropriate. Sometimes a research question has independent variables that cannot be manipulated. One of the most effective uses of ex post facto research is for future experimental manipulation; ex post facto research can help identify a small set of variables within a large set of variables related to the dependent variables.

So far, the discussion of threats pertains only to internal validity and is derived from the original work by Campbell and Stanley (1963) and the more recent guidelines of Shadish et al. (2002); we now move to external validity, which addresses concerns for generalizability. Shadish et al. (2002) categorize these threats as conceptually similar to statistical interactions: "It is the concept behind the interaction that is important—the search for ways in which a causal relationship might or might not change over persons, settings, treatments, and outcomes" (p. 86).

Threats to External Validity

Those authors illustrate five threats to external validity, including the interaction of causal relationships and (1) units being measured (perhaps the results apply to men but not to women), (2) various forms of the treatment, (3) various forms of the outcome measure, (4) different settings, and (5) the context specificity (the process by which the cause transfers into an effect possibly operating differently in different contexts).

Several threats to external validity, according to Campbell and Stanley (1963), include:

- First, the reactive effects of testing in which the pretest might cause the participants in the study to be either more or less sensitive to the treatment.
- Second, the reactive effects of the research situation itself threaten generalizability (in both cases, the sample of participants and their experiences might be unlike the population to which the researcher wants to generalize).

- Third, generalizability can be lessened when there is interaction between biases in the selection of participants and the treatment.
- Lastly, multiple-treatment interference can limit external validity as well. That research participants are subjected to more than one treatment can result in prior treatments having an effect on later treatments. The results of the later treatments cannot be generalized to groups who have not experienced the earlier treatments.

Replicability

One way of thinking about external validity is replicability, the extent to which the study can be repeated in the same or in a different system (Newman, McNeil, & Fraas, 2004). The importance of replicability has been largely ignored because the problem has been misunderstood. Frequently, replicability has been confused with statistical significance.

Statistical significance and replicability are related in some sense, but the relationship is not a linear one. Newman et al. (2004) suggest that the alpha levels traditionally used in education may be too high, as the following example illustrates. In a study with an N of approximately 12, a researcher might find a statistically significant finding at the .05 level, meaning one is 95% confident that the difference was not due to chance in the population. However, the replicability may be as low as 50%, not even near 95%. On the other hand, if the difference is found to be significant at the .01 level, the replicability may be as high as 72%. And, further, if one finds statistical significance at the .001 level, the replicability could be as high as 90%. We need to recognize that because a finding is significant at .05 half of the time, it is not likely to replicate. Education researchers may need to consider paying more attention to replicability estimates as the "standard" for accepting a research outcome as meaningful, rather than statistical significance alone. A more thoughtful interpretation of statistical findings will result, making implications much more cogent to audiences. Perhaps these insights will lead researchers to set their alpha levels much lower than the traditional .05 (Newman et al., 2004).

Multivariate Research

Before moving from the quantitative paradigm to similar issues in the qualitative paradigm, a brief mention of *multivariate research* is needed.

This segue into qualitative research is appropriate because rather than investigating one variable at a time, multivariate research designs, appropriately, address questions of the human, organizational, and social complexities of everyday life. In these designs, multiple dependent variables are analyzed simultaneously.

More than one variable is usually operating in the questions researchers ask. In the qualitative research situation, human complexity, at some level, underlies the research question. Researchers are not driven by few variables but rather are exploring phenomena more holistically and in unbounded ways. Multivariate research goes some distance to appreciate that complexity within the quantitative paradigm.

Quantitative researchers may use single dependent variables (*univariate design*) or multiple dependent variables (*multivariate design*). Univariate design's one dependent variable (or criterion) is being predicted by another set of one or more variables (independent or predictor variables). In multivariate design, two or more dependent (or criterion) variables are being predicted by two or more independent (predictor) variables. Common complexities of social phenomena lead to research questions with many variables. Higher-order factorial designs (two-way, three-way, four-way, and the like) more closely approximate the external world than do one-factor designs.[2] In essence, each additional independent variable increases the design's relevance to reality as long as such variables account for some unique variance in the dependent variable. For example, the effects on math achievement are not explained merely by IQ but more credibly by IQ, family background, parents' education, and socioeconomic status. However, with each additional variable, the number of cases in each cell may decrease (the participant-to-variable ratio decreases). As the number of variables increases, one must consider increasing the sample size. Therefore, some researchers use multivariate analysis as opposed to univariate analysis to increase external validity. In other words, behavioral phenomena usually have an impact upon more than one dependent variable; measuring one while accounting for the variance of others is more likely to result in conclusions that reflect relevance to the external world.

According to Patton (2002), qualitative methodology is based on the assumption that the study of human behavior must be different from the study of nonhuman phenomena. Strike (as cited in Patton, 2002) claims that "human actions are intelligible in ways that the behavior of nonhuman objects is not" (p. 52). In short, a human being lives in a world that has meaning; and, because one's experiences have meaning, that meaning can be discovered and explained (Bar-On & Parker, 2000).

While Patton argues for the qualitative paradigm, this view is not so different from a stance taken by some quantitative researchers. In attempting to try to understand even the simplest human experience, quantitative researchers claim that one must examine many variables and the situations in which they occur. In other words, variables may have differential effects, depending upon the specifics of the situation. They point to design strategies that control for multiple effects on dependent measures. Following Patton, one could test for the interactive effects of emotion and purpose on a certain behavior. Furthermore, "perceived ability," "likelihood of failure," and "need states" could be measured, and their relationships to behavior could be tested. Interaction effects are more reflective of real world behavior than are simple effects (McNeil, Newman, & Kelly, 1996; Raudenbush & Bryk, 2002; Shadish et al., 2002). Although Patton implies that science can relate only to main-effects questions, one can use advanced scientific design methods that investigate complex human experiences.

Validity in Qualitative Research

In a 1995 article, Lincoln chronicles the discussions about validity in naturalistic research. She places in historical context the early conceptualization of criteria she and Guba first described in 1985. In the final section of this chapter we present their criteria—after discussing several other writers' notions of validity since 1985.

Validity is the truth value of a research study, and, therefore, a central concern for all researchers. While Denzin's and Lincoln's (2005) comments about validity as a "hot" topic set the stage for discussions about validity in qualitative research, in reality no consensus about validity in qualitative research yet exists (Toma, 2006). But it remains important, according to Onwuegbuzie and Daniel (2003). They found that one of the most common errors of published qualitative research studies was lack of validity evidence.

As noted in our earlier discussion of the philosophical underpinnings of qualitative and quantitative research, the idea of *truth* is problematic because truth becomes a social construct, idiosyncratic and situationally specific. Is validity even possible given the nature of qualitative inquiry? In her article, Lincoln outlines the responses to the "crisis of legitimation" in qualitative research (Denzin & Lincoln, 1994). Many methodologists in the field of qualitative research have responded to this question of validity. Next we briefly discuss some of these authors' ideas about validity to put our response, the postpositivist perspective, in context, and eventually, we return to Lincoln and adopt criteria she and Guba first articulated.

Hammersley (1992) describes four philosophies or stances that call for different ways of thinking about qualitative research. First, he describes "methodism," his synonym for positivism. Within this philosophy, the scientific method is a way to truth. Hammersley contends that in this view we cannot know anything that exists outside our experience. Validity cannot be connected to an external reality, so it is connected to what we usually think of as reliability defined as "(agreement between the findings of different observers or between the findings of the same researcher on different occasions) and/or predictive validity ([defined as] agreement between the results of the research and established measures of the relevant property)" (p. 196). He contrasts methodism and "realism." In this second philosophy, the researcher establishes validity because the researcher gets close to the participant under study. There is a strong correspondence between the knowledge gained in the research and the reality it represents. This philosophy assumes a reality out there that can be known. Third, because ethnographers construct the reality of the people or cultures they study, this research requires validity that he calls "relativism." Relativism accepts multiple valid accounts of reality based on this constructivist epistemology.

Page (1997) also suggests that validity is a social construction in qualitative research—"a judgment produced in the relationships established between the author of a text, her participants, and readers. It can change over time or with new information" (p. 151). She argues that the interpreter needs to create a "coherent interpretive framework" (p. 151) in order for authentic understanding to be conveyed (by the researcher) and achieved (by the reader). The researcher's responsibility is to provide some sense of boundaries within which meaning can be known—because any truth is necessarily partial and only one of multiple potential truths. Her stance is similar to the discussion of the nomological network, that is, within that network, one truth may be supported by data, and another truth may not be supported.

Consensus within a community about the truth of the findings constitutes validity in these cases described by both Hammersley and Page. "Instrumentalism," Hammersely's fourth philosophy, abandons validity concerns and opts, instead, for research that does some good, similar to the purposes of critical theorists and feminists.

As Hammersley moves within several notions of validity, he questions the construct of truth and validity in a postmodern world. Multiple realities are acknowledged, and in a postmodern context, Tierney (1993) uses ethnographic fiction to explore organizational life. He cites Kitzinger to define the purposes of such a research strategy: "The attempt to induce in the reader a willing suspension of disbelief, while simultaneously acknowledging that one's argument is an 'account,' a 'construction,' or 'version,' rather than objective truth can take social constructionist writing into the realm of the arts" (p. 313). Tierney comments, "Through fiction, then, we rearrange facts, events, and identities in order to draw the reader into the story in a way that enables deeper understandings of individuals, organizations, or the events themselves" (p. 313).

Given this purpose and this method, validity assumes quite a different perspective. Tierney posits that validity then asks questions: Are the characters believable? Is this situation plausible? Has the text led me to reflect on my own life?

Lather (1995) poses "four post-modernist kinds of validity" as alternative ways of thinking about truth that reject correspondence theories of truth (positivism) (p. 54). "Ironic validity," "paralogical validity," "rhizomatic validity," and "embodied validity" question the "truth" for postmodernists.

Lather's *ironic validity* fits the postmodernist epistemology because it considers truth a problem. The truth value of the research lies in its ability to show coexisting binaries, coexisting opposites (p. 57). *Paralogical validity* is that quality of research that legitimates because it reveals paradoxes, "undecidables," parts of meaning that are incapable of being categorized. Lather describes the legitimation of research that includes excerpts from interviews that are unmediated by the researcher, that cannot be interpreted, or that are not interpreted because of the researcher's unwillingness to diminish them (p. 57). *Rhizomatic validity*, unlike the deeply ingrained meanings that solidify the postpositivists' categories through content analysis, fits the postmodern rejection of stable truth. This validity comes through "the crossings, overlaps, the meanings with no deep roots" (p. 58). (This is evocative of path analysis in the quantitative paradigm: following where the data patterns take one in constructing a theory.) *Embodied validity*—validity that comes from the idiosyncratic nature of the study—comes from what Lather describes as the researcher knowing more than he or she is able to know, writing more than he or she is able to understand (p. 59). In essence, this is the nature of interpretation, bringing a sort of closure to the intellectual and emotional sorting and sifting of data.

Because the knowledge generated by qualitative research is the result of a social construction, Kvale (1995) claims this phenomenon is related to construct validity and is the goal of qualitative research. He presents three approaches to validity generated by this reasoning. The correspondence theory of truth does not apply in this postmodernist view. The social construction of reality is validated only through practice.

Validity as a concept seems to imply a boundary line between truth and nontruth, according to Kvale. While there is not a universal truth, there may be "specific, local, personal, and community forms of truth, with a focus on daily life and local narrative" (1995, p. 21). More clearly than other writers, he explains the postmodern perspective.

> The postmodern condition is characterized by a loss of belief in an objective world and incredulity toward metanarratives of legitimation (Lyotard, 1984). With a delegitimation of global systems of thought there is no foundation to secure a universal and objective reality. The modern dichotomy of an objective reality distinct from subjective images is breaking down and is being replaced by a hyperreality of self-referential signs. There is a critique of the modernist search for foundational forms and its belief in a linear progress through more knowledge. The dichotomy of universal social laws and unique individual selves is replaced by the interaction of local networks, where the self becomes an ensemble of relations. The focus is on local context and on the social and linguistic

construction of a perspectival reality where knowledge is validated through practice. (p. 24)

Kvale refers to the notions of validity in Polkinghorne: "Validation becomes the issue of choosing among competing and falsifiable interpretations, of examining and providing arguments for the relative credibility of alternative knowledge claims" (Polkinghorne, 1983, p. 26). Kvale labels his validity constructs *investigation validity*, *communicative validity*, and *action validity*.

Investigation validity is the quality of craftsmanship. Quality of one's work and the appropriateness of tools are certainly related, but in addressing validity, they are best kept as two separate issues, conceptually similar to Lincoln and Guba's (2000) call for authenticity checks and Toma's (2006) call for rigor. A type of validity uniquely applied to each study, this validity is impossible to standardize. Here, the concern is quality, not the match among the research purpose, question, and method (what we propose as the heart of validity). The focus is not on the method itself but rather how well the method was carried out by the researcher.

Kvale lists checks that the researcher should make after interviewing, for example. The consistency of what a participant says in an interview is checked against other statements he or she makes. Other participants are interviewed, for example, in a kind of triangulation. Investigation validity includes how theories are derived from the data. For example, Kvale discusses research on student grades. Does the theory emerging from the data relate to assessing knowledge, or does it relate to the political dynamics of separating groups of students? To the extent that the theory aligns with the data and the purpose, one assumes validity (Ridenour et al., 2003).

Kvale's second validity is what he calls *communicative validity*.

> Communicative validity involves testing the validity of knowledge claims in a dialogue. Valid knowledge is not merely obtained by approximations to a given social reality; it involves conversation about the social reality: What is a valid observation is decided through the argumentation of the participants in a discourse. (p. 30)

His third validity is *action validity*, in which truth is justified based on whether or not it works. Kvale refers to Patton (1990) and Patton's concept of credibility and likens it to his own term of action validity. According to Patton, the test of credibility of an evaluation report is whether it is used by decision makers. This is similar to the focus on pragmatism by Tashakkori and Teddlie (2003) as an argument for mixed methods designs: "One major reason [for pragmatism] is that mixed methods are often employed in applied settings where practical decisions stress the utility of multiple data sources for decision-making purposes" (p. 679).

Truth, according to Kvale, is whatever is helpful in taking action to get a desired result. He makes comparisons to Polkinghorne (1991), who claims that the validity of case studies, narratives, and so on can be tested against their effects on practice. Kvale's final reference is to Lincoln and Guba (1985), who say that inquiry helps understanding, and through understanding, participants increase the control of their lives. For instance, when one asks a set of questions, the activity forces them to be reflective. Questions about validity force participants to be reflective, and that alone moves them closer to Winter's (2000) claims of validity—the correlation of research methods and the purposes of the research, rather than any standardized procedure (p. 11).

In their 1985 classic work, *Naturalistic Inquiry,* Lincoln and Guba suggested strategies to enhance truth value of qualitative research. We adopted these for our model of the validity of qualitative research; we attribute the following list entirely to them. Others have built on their criteria as well (e.g., see Goetz & LeCompte, 1984; McMillan & James, 1992; Toma, 2006).

Guba and Lincoln (2005) claim the 1985 criteria are still useful, but they more recently theorize truth value as both an ethical issue and embedded within expectations of strong research rigor. And rigor, they claim, is a dual concern: rigor in the application of methods as well as in the interpretation of data:

> Prior to our understanding that there were, indeed, two forms of rigor, we assembled a set of methodological criteria, largely borrowed from an earlier generation of thoughtful anthropological and sociological methodological theorists. Those methodological criteria are still useful for a variety of reasons, not the least of which is that they ensure that such issues as prolonged engagement and persistent observation are attended to with some seriousness. (p. 205)

Design Validity Criteria (Predominantly Qualitative)

What follows is Guba and Lincoln's list of criteria that help one begin analyzing predominantly qualitative research for its design validity. Questions that might be used to probe the validity of methods in predominantly qualitative studies are included. These strategies are suggested as beginning points—not as an exhaustive list.

Prolonged Engagement On-Site

The purpose of prolonged engagement is to detect trends. Did the researcher observe long enough to get an accurate reflection of the culture or history? If only one observation is taken, it will reflect one small portion and will not capture the essence of the culture or the situation. For instance, at a recent American Educational Research Association (AERA) meeting, David Berliner told the story of observing a teacher's classroom because his wife claimed that this teacher was the best she had ever seen. To his surprise, the children were running all over the classroom, chaos seemed to reign. Later, he learned that three circumstances had intervened immediately prior to his visit: it had rained all day, it was the period after recess, and the children had not had a break. What he witnessed was not a sufficient picture of the classroom in its entirety. He had seen a unique situation.

Persistent (Consistent) Observation

Was sufficient time spent on-site to get a full and consistent portrayal of behavior? Was the observation typical or something that rarely occurs? The research report should reflect this. While the purpose of prolonged engagement is to be able to detect cultural trends or idiosyncrasies, the purpose of persistent observation is to identify or estimate if a particular behavior is (or sets of behaviors are) frequent or infrequent.

Triangulation

Did the researcher attempt to obtain a variety of data sources (e.g., different observers or different written histories)? If so, was there shared reality? Although an important concept in much of the qualitative methodological literature, it might be considered somewhat quantitative in that the researcher is looking for consistencies in perceptions. On one hand, triangulation might be looked at as a reliability check.

On the other hand, it is possible that one source of data is more important than other sources in understanding a particular phenomenon. Generally, however, the more sources a researcher uses the more likely he or she is to have a full and rich interpretation.

Peer Debriefing

Did the researcher talk with any other professionals to get another perspective on what he or she saw or experienced? For example, a researcher may get attached to the people and the setting under study. Researchers may begin to interpret things from their own need base. They need to check out their emerging construction of meaning with other professionals or colleagues.

Negative-Case Analysis

Has the researcher taken into account all known cases; and has he or she continually revised the emerging hypothesis until all known data are explained by the hypothesis? In essence, the researcher expands and reshapes interpretation until all outliers are included.

Referential Materials

Did the researcher use enough supportive material (e.g., documented recordings, readings, archives, or other materials that are available to others)? It is important for the researcher to document references, records, and interviews used and to let readers in on how knowledgeable the researcher is of these materials. It is also important to indicate which sources were used in what ways and, if available sources were not used, why they were not.

Member Checking

Member checking refers to how accurate the data are. Were the data and interpretations continuously checked? One way of estimating the accuracy of personal observations is to check out those observations on members of the group under observation, i.e., to return to and question those people interviewed to make sure that he or she got it right.

Thick Description

Lincoln and Guba (1985) strongly suggest that the researcher include the "widest possible range of information" in their research report (p.

316). The reader should be able to comprehend the full portrait of the research setting because of the rich details in the account. With his or her "story," the researcher wants to place the reader as much as possible within the scenario of the study.

Leaving an Audit Trail

Does the researcher have good documentation so that another researcher can easily replicate the research? This not only means that someone would be able to replicate the current study but be able to confirm or to contradict the interpretation based on the same data.

Reflexive Journal

Did the researcher maintain a daily journal in which he or she writes about the events of the study that day, personal reflections on how his or her values and beliefs came into play during those events, and, lastly, decisions about methods as they continued to unfold? Lincoln and Guba (1985) refer to this as "reflexive" as it documents the dynamics of the "self," the "instrument" of data collection and analysis in qualitative research.

Theoretical Sampling

Did the researcher follow where the data led? The researcher typically enters the field and begins immediately to collect data. While data are being gathered, the investigator has begun to form explanations of their meaning. These tentative explanations (theory) suggest other data sources. In other words, the sampling of data in qualitative research is determined by the existing data (Goetz & LeCompte, 1984). The researcher attempts to capture the best theory that explains the data. A quantitative researcher might call this *soft hypothesis testing*. The researcher may change direction or collect different or additional data. On the other hand, such data sampling may provide supportive and corroborative interpretations of initial emerging theory.

Structural Relationships

Is there logical consistency among different data sets? The researcher should construct meaning as much as possible by interweaving different data sets, which may come from different perspectives, while supporting common underlying and emerging meaning. Another way to conceptualize this is to develop a nomological network (Cronbach & Meehl, 1955; MacCorquodale & Meehl, 1948). A nomological network is a set of propositions that have related logical meaning. In other words, the researcher might identify constructs (or "laws" in Cronbach and Meehl's term) and how one would expect them to be related logically. These constructs may come from observables ("data") that are then related to theoretical constructs, theoretical propositions that are related to one another, or any combination of patterns between observables and theoreticals. For example, a school in which data suggest a sense of trust between teachers and administrators would be expected to also show a positive school climate. The constructs of "trust" and "school climate" would logically be related. Multiple sources of evidence are suggested to the researcher as he or she establishes these structural relationships.

Generalizability

That one should be able to generalize underlies science. If the purpose of the research is to generalize, we assume one should employ quantitative methodology. Not all researchers would agree; and generalizability is often conceptualized differently in qualitative methods.

Polkinghorne (1991) distinguishes between two types of generalizability: *statistical* and *aggregate*. The statistical model is more consistent with quantitative assumptions, and the aggregate model, based upon deep description, is more consistent with qualitative assumptions. A deep and rich description is sufficiently comprehensive to allow the qualitative researcher to generalize to each member of the population. Donmoyer (1990), too, states that generalizability can be broader than traditionally defined. He accepts traditional generalizability for statistical research (quantitative) and develops schema theory based on Piaget's notion of *assimilation, accommodation, integration,* and *differentiation* (qualitative research) (p. 197). The growing body of published qualitative research has put pressure on research methodologists to create ways in which results of such studies can be applied to wider audiences, or generalized. The concepts of *applicability, transferability* (context limited), and *replicability* are examples of the kinds of questions to ask to improve those efforts.

Applicability. Can this research be applied to other samples? The reader must look at the sample size as well as the sample characteristics. Important assumptions are that the hypothesis is an emerging one and not a sample-to-population statement, that there is no test of statistical significance, and that the purpose of deep description is to describe in detail the characteristics of the sample being investigated so that others can make logical judgments about whether the sample is comparable to other samples.

Transferability. Do the findings of the research hold up in other settings or situations? To the extent that it can be argued from a logical or a data-based point of view that what is being observed is not context limited, then the findings are transferable to other contexts and therefore generalizable. For example, the effects of praise on a student may be more context independent than context specific. Praise may affect academic performance as well as performance in sports.

Replicability. What is the likelihood that a given outcome or event will happen again if given the same circumstances? Replicability is difficult to accomplish with any level of confidence, especially outcomes of a naturalistic study. One must identify changes that are due to identified effects and the frequency of these common occurrences at different points in time, in different settings, and by different observers. When these data are available, they can provide important insights.

Truth Value (Credibility)

What confidence does the reader have in the findings of the research? We previously described the concept of construct validity, claiming that an instrument has construct validity to the extent that it has all other types of validity (i.e., face, content, expert judge, concurrent, and predictive). Similarly, a study has truth value to the extent that the above qualities or components are built into the design. Any one study is unlikely to have all qualities. Some of these components are more important than others, in certain studies; and, generally, the more components the greater the truth value.

Validity in Mixed Methods Research

Mixed methods designs can produce valid research findings, especially the qualitative-quantitative interactive-continuum model of mixed methods. First, choosing from multiple "ways of knowing" is the researcher's task. Conflicting ways of knowing about the world and different epistemological points of view have been continually debated. Is reality "constructed" by the researcher, one who is a subjective insider intimately involved in interpretation and acting as the "tool of data analysis"? Or, is reality "out there," to be captured objectively by the researcher, an outsider and detached knower? When a research question and purpose call for both qualitative and quantitative methods, the researcher needs to choose methods and align them to purposes and research questions in a legitimate mixed methods approach. Tashakkori and Teddlie (1998) resolve these questions about ways of knowing and truth value for themselves by resorting to a type of pragmatism that they call "what works."

Rather than focus on the dilemma posed by conflicting ways of "knowing," they conclude that qualitative and quantitative methods are "compatible." They express a faith in the belief that sufficient commonalities exist *across the qualitative and quantitative divide* that a third methodology, mixed methods, provides a salient bridge.

> It can be argued that there is a common set of beliefs that many social and behavioral scientists have that undergird a paradigm distinct from positivism or postpositivism or constructivism, which has been labeled pragmatism. This paradigm allows for the use of mixed methods in social and behavioral research. (p. 13)

The "common set of beliefs" is an effective way to capture the fundamental purposes of all researchers, regardless of the paradigm within which they work. These common beliefs should include the notion of science and include the obligation of the researcher to justify design validity. On the other hand, an alternative argument is that researchers cannot so easily ignore issues of epistemology and truth by merely assuming a compatible middle ground. Pragmatism can open the door to the easy compromise, the quick choice of methods based on the criterion of "what works" rather than on the criterion of the purpose of the research. It is the latter, the research purpose, along with the research question, that must determine the choice of methods (Newman et al., 2003).

Others conceptualize mixed methods. Johnson and Turner (2003) define the purpose of mixed methods: "[M]methods should be mixed in a way that has complementary strengths and nonoverlapping weaknesses. . . . [The principle of mixed methods research] involves the recognition that all methods have their limitations as well as their strengths" (p. 299). Here, Johnson and Turner echo the concerns that Brewer and Hunter (1989) articulated over a decade earlier, that is, that "complementary methods" allow one set of methods to make up for the "flaws" in other methods. Whether or not this complementarity can be achieved is worthy of discussion; how qualitative and quantitative methods can "complement" each other is idiosyncratic to each research situation. For instance, a researcher's question may ask how community groups differ in their approaches to retention policies in their elementary schools. The researcher's purpose is community satisfaction with changed policy and so creates a questionnaire addressing retention and uses it to gather numerical ratings on twenty-five items from three hundred citizens, randomly selected samples of adults distributed throughout the census tracts of the school district. With over 80% responses, the researcher feels confident he or she can represent the profile of the community's perceptions.

The twenty-five-item questionnaire, however, produces data only on those items the researcher knew about before the study began. He or she decides, therefore, to select a sample of sample individuals in the community to delve into their experiences, understandings, and feelings about retention through in-depth interviews. This strategy will give the researcher new data—novel understandings from the community that might not have been anticipated.

The quantitative and qualitative strategies fulfill different purposes; they are not flawed for not fulfilling *every* need of the researchers. A questionnaire strategy requires that the researcher know the structure and potential content of all the answers before the study begins; an in-depth interview does not. The in-depth interview requires the researcher to expend extensive amounts of time in collecting data from very few individuals, the antithesis of the questionnaire. These are not flaws. The efficiency of collecting the same data from many people through a questionnaire is a benefit, an advantage, a strategy that is consistent with the researcher's need. That the interview represents only a relatively few citizens is not a flaw, the richness of new and unanticipated understandings that are achieved is a benefit.

Methods can only be judged in relation to particular research purposes and questions (Newman et al., 2003; Sandelowski, 2003). For example, that randomly selected statistical samples are not "thick" and "rich," according to Sandelowski (2003), is not a weakness or a limitation of quantitative methods. Similarly, that a case study cannot produce generalizations is not a weakness of qualitative methods. "Rather," Sandelowski claims, "it is the researcher who is weak or limited who chooses inquiry approaches for the wrong reasons, executes them in the wrong way, or apologizes for method characteristics that require no apology" (p. 329). Methods can play different roles in providing evidence: they have different functions. But these cannot be judged strong or weak absent an actual research context.

Researchers should select research methods based on the purposes of the research study and on their research questions. It is only within the context of the research purpose that any methods can be judged "weak" or "strong." These judgments are based on whether they fulfill the purpose of the study and the research questions that necessarily must drive the study.

When the researcher can justify that the methods he or she has selected do indeed fulfill the purpose and address the question, then the methods are consistent with those purposes and questions (Newman et al., 2003). By definition, then, they are "strong" (not "weak") for that study. For example, such a researcher would *not* select a case-study method when the purpose of the study was to test group differences in math-achievement scores after a pedagogical intervention (Newman et al., 2003). Rather, the purpose would be fulfilled by selecting a random sample of participants, assigning them randomly to treatment groups, collecting valid and reliable pretest and posttest data on selected variables, and statistically analyzing the relationships to test hypothesized group differences. Furthermore, *adding* a case study to this design in order to strengthen the so-called "weakness" of statistical hypothesis testing would be of absolutely no value at all. Only if the case study served an authentic purpose would the study be strengthened. Legitimate, strong, and scientific research studies can be designed *solely* to fulfill the purpose of testing hypothesized group differences. No apologies are required of researchers who carry out hypothesis-testing quantitative studies. Of course, parallel examples could be offered for a predominantly qualitative study. Because epistemological assumptions underlie all of what we can claim to "know" and because the researcher is involved in systematic "knowing," then these epistemological assumptions need to be central to the researcher's thinking.

Epistemological Assumptions Underlie All Scientific Research

Education researchers are scientists, and, as such, they must grasp the epistemological underpinnings of their research studies. To achieve the qualities of systematic processes and thinking, potential verifiability, potential replicability, self-correction, and explanation of phenomena (five criteria of science presented in chapter 1), the researcher's assumptions must be clarified and consistent with the purpose of the study and the question being posed (Newman et al., 2003). On the one hand, mixed methods research *thoughtlessly* done can weaken appreciation of underlying assumptions. On the other hand, mixed methods when *thoughtfully* done can strengthen appreciation of underlying assumptions. Assuming that one is free from justifying any knowledge claims about his or her evidence weakens the scientific quality of the research. Being liberated from having to justify knowledge claims because one is using "complementary" research paradigms and so has all epistemological bases covered might constitute a pitfall of mixed methods if they are selected without serious thoughtfulness and planning.

Consistency—the Validity Criterion

We argue that the criterion of consistency is the one to which the researcher should aspire to achieve validity. Consideration of methods themselves as strong or weak is not helpful to the researcher. Consistency among purpose, question, and design is the standard criterion for planning studies of high quality and scientific value.

This notion of consistency as validity shows how mixed methods designs can be aligned with the five criteria of science. One can mix methods and philosophies to address different components of the same study. For example, a researcher can identify barriers that need to be overcome in solving a problem in a particular setting through focus groups (qualitative methods and qualitative assumptions). Then, the researcher can use the emergent themes from the transcripts to develop operational definitions of barriers that can be measured as quantifiable variables. The researcher can assess the extent to which the identified solutions are effective with a sample of participants, results that may be generalized to a larger population (quantitative methods and quantitative assumptions). This mixed method uses one method (qualitative) to inform the other (quantitative). This is an example of one category of mixed methods designs, one category of the three that we perceive in the literature.

Strategies to Enhance Validity and Trustworthiness

Research design is made up of the methods one selects to carry out the study. The methods implement the design—the focus of chapter 3. The discussion of legitimacy of design continues by focusing in this chapter on the qualitative paradigm. Although ways to mitigate threats to the validity of quantitative research are well recognized, ways to mitigate threats to qualitative research are not universally accepted (Toma, 2006). From our postpositivist perspective, we present ways to enhance the design validity of studies.

This chapter describes ways to

- improve the validity of observational methods
- improve the validity of grounded-theory methods
- improve the validity of case-study methods
- improve the validity of interviewing methods
- improve the validity of historical methods
- improve the validity of ethnographic research
- improve the validity of phenomenological research
- configure triangulation and its effects on truth value

Observational Methods

Observation is a frequent data-collection method used in qualitative research. Observation is frequently categorized in three ways: (1) Participant observation (in which the observer is obvious to and involved with the participants) creates a situation in which the researcher and the participants develop rapport and naturally and comfortably interact with one another over time, (2) Reactive observation is useful when participants are aware of being observed in a setting where the researcher controls the interactions, and (3) Unobtrusive observations occur when participants are unaware of being observed (Angrosino, 2005).

Gay (1987) and Mouly (1970) discuss the potential invalidity of observational data when they call for, at the very least, a scientific basis for the observation. Mouly, more specifically, agrees that both the scientist and the layperson observe, "but the scientist starts with a hypothesis and arranges the conditions of his observations to avoid distortions" (p. 282). He warns further about invalidity, especially of participant-observation techniques.

> As the participant observer adapts more and more to his role as a participating member of the group, he becomes increasingly blinded to the peculiarities he is supposed to observe. As a result, he is less likely to note what would be significant to a more objective observer. As he develops friendships with the members of the group, he is also likely to lose his objectivity, and, along with it, his accuracy in rating things as they are. (p. 289)

Despite these pitfalls, there is validity in using the observational method for study of some phenomena, such as nonverbal behaviors.

All validity concerns described here affect participant, reactive, and unobtrusive observations. In participant and reactive observation, the researcher is a regular participant in the activities being observed; in unobtrusive observations, the researcher is not a participant in the ongoing activities being observed. Compared to *participant-observation* strategies, the validity of *unobtrusive* strategies is greater because there is no reactivity among the participants to the presence of the researcher. This reduction in bias, however, does not cancel out the other biasing (invalidating) effects.

Despite these limits on the validity of observational methods, some maintain that it is, nevertheless, a highly appropriate technique (e.g., Hakim, 1987). Lofland (1971), for one, designates a first priority to the observer's understanding of the participant's point of view: "In order to capture participants 'in their own terms,' one must learn their categories for rendering explicable and coherent the flux of raw reality. That, indeed, is the first principle of qualitative analysis" (p. 7).

While this "understanding of the participant's point of view" is highly regarded, the statements may well be describing only observer bias. However, Becker and Geer (1960), as well as Lombard (1991) and Lincoln and Guba (1985), place the methodology in even higher esteem when they state that participant observation is the "most comprehensive of all types of research strategies."

> The most complete form of the sociological datum, after all, is the form in which the participant observer gathers it: an observation of some social event, the events which precede and follow it, and explanations of its meaning by participants and spectators, before, during, and after its occurrence. Such a datum gives us more information about the event under study than data gathered by any other sociological method. (p. 133)

The observer's attention to a setting is described as an evolving role by Boostrom (1994). From his own experience, he shows how the qualitative researcher can move from an "almost inert receiver of visual and aural stimuli" to being interactive in constructing the account of what he or she sees. He sees this role change as a move through the roles of "videocamera, playgoer, evaluator, subjective inquirer, insider, and, finally, reflective interpreter" (p. 53).

Enhancing Validity of Observational Methods

Observers construct meaning from what are usually data that are "objectively" recorded in a well-intended manner. However, objectivity can be elusive. Because the researcher is the instrument of data collection and analysis, full disclosure of the researcher's lens—identifying predispositions toward interpreting meaning from the observations because of race, class, gender, or any number of characteristics lends validity to the report of observational results, according to Angrosino (2005). Furthermore, Angrosino (2005) suggests that standardized procedures are impossible when applying observation methods—the strategies are idiosyncratic to the setting. With increasing contemporary focus on the intimate interaction between the researcher and the participants, he suggests the principle of "proportionate reason" as a guide to observers. To determine whether the proper interaction exists among the participants and the researcher, he offers that the researcher first assess whether inserting himself/herself into a setting harms those in that setting. Second, researchers must select the observational strategy that results in the least amount of cost to the participants' lives; and, third, the methods of observation do not compromise the value of the observation.

Volunteering as a classroom tutor in a program that serves adults with mental retardation whom one is interested in observing and interviewing is probably sufficiently proportionate; in contrast, becoming a bill-paying benefactor to induce cooperation among such adults in a group home would be morally questionable. (p. 736)

In addition to these broad decision guidelines, six strategies can strengthen validity. First, because each observer brings a complex set of identities, several observers from several backgrounds (or points of view) reporting on the same phenomena can increase validity. Coalescing their data can reduce sensory-deficiency and misinterpretation error. Second, Angrosino (2005) also suggests that researchers focus on the particular, not the general. Third, structuring the observation (a reactive observation) increases validity by focusing the attention of the observers on certain characteristics and events. Fourth, placing the observation on a scientific foundation by stating a hypothesis upfront increases validity by avoiding distortion. Fifth, unobtrusive observation, as opposed to participant or reactive observation, increases validity. And, sixth, using observation only for studying those phenomena that are appropriate to this method (e.g., nonverbal behaviors and social interactions) increases validity.

Grounded-Theory Methods

Grounded theory is both a type of design as well as a data-analysis strategy. Observation is the first and key data-collection strategy of the grounded theorists. To relate validity concerns to the grounded-theory methodology, we first need to review the approach, described in the classic work by Glaser and Strauss (1967).

The observers enter the research situation with no hypothesis. They describe what goes on, and from the observational data, they develop a hypothesis. Charmaz (2005) elaborates:

Essentially, grounded-theory methods are a set of flexible analytic guidelines that enable researchers to focus their data collection and to build inductive middle-range theories through successive levels of data analysis and conceptual development.... Grounded-theory methods consist of simultaneous data collection and analysis, with each informing and focusing the other throughout the research process. (p. 507–8)

Grounded theorists simultaneously address the process of research and the product of research; these processes are inseparable. As information emerges from the data, it is put into an original theoretical framework.

Grounded theorists do not use traditional quantitative methods to verify the data. They move back and forth from data collection to analysis, letting the emerging tentative analysis lead to new sources of data and as a way to check developing ideas (Charmaz, 2005).

Data are analyzed by initial and focused coding techniques. *Initial codes*, constructed after studying the data, are derived from the participants and the roles they play, the context, the timing and structuring of events, and the issues that are the focus of the participants' behaviors and interactions. Connections among all these phenomena are attended to as the researcher codes the data.

The next phase consists of *focused coding*. Here the initial codes (labels) are combined and raised to an analytical level and put into categories. Sometimes, categories are developed in the language of the participants, and sometimes the researcher devises categories in his or her own terms. Categories can be broken up and recombined. The researcher may go to the literature to expand and clarify the codes and categories. In these instances, researchers use the literature as "a source of questions and comparisons rather than as a measure of truth" (Charmaz, 1983, p. 117). Grounded theory emphasizes process, and categories, once developed, are not treated singly but are woven together to make meaning. This is a processual analysis.

Charmaz (2005) characterizes contemporary twenty-first-century grounded theory as a departure from the more positivist traditions established by Glaser and a move toward a more constructivist, less objectivist, set of assumptions. Charmaz claims that more recent advocates of grounded theory focus on how the researcher is located within the studied phenomena, and they do not assume that observers are naïve blank slates.

Enhancing the Validity of Grounded-Theory Methods

Data collected and results formed in grounded theory can be made more valid by use of the alterations to observational techniques discussed earlier. It is immediately apparent, however, that beginning with a hypothesis at the outset of data collection violates the first and most important assumption of this method. Yet, the initial-coding stage and the focused-coding stage of the process are not unlike an empirical researcher's coding of open-ended questions. This similarity reveals an area of overlap in qualitative and quantitative methods. A common process for the empirical researcher is to collect responses on a questionnaire, review all responses to a particular inquiry, and then categorize them. The resulting categories are used as categorical variables in a statistical analysis. This traditional approach does not include, however; the ambiguous process of sifting and resifting, considering what information was left out, what was said, and other subjective manipulations of responses. The processes described in the Charmaz chapter imply a highly skilled, insightful coder who has an unusually profound understanding of the psychology of human motivation, incentive, and behavior.

The theory-building purpose of the grounded-theory method rests on underlying assumptions about procedures that are highly questionable, considering the definitions of validity assumed here. Therefore, initial coding may be misleading, and the resulting theory based upon the coding may be unreliable.

Grounded-theory methods as represented here in brief are not without validity concerns. To begin, no hypothesis directs the data collection. As with unobtrusive observations and unstructured interviews, validity might be diminished when the researcher is potentially bound only by his or her biases. Because the grounded theorists maintain an "independent" view of the data, untainted by even a literature review, it seems difficult to accept that their perceptions are bias free. When the researchers do not acknowledge factual assumptions at the outset, which is applauded as freeing the researchers to experience solely the constructs emerging from the data, they may be allowing their subjective perceptions full rein. That they go so far as to acknowledge that the data they collect may not even be related to the topic under study is a clear indication of the design's potential for invalidity.

Collecting "data" in a grounded-theory study requires researchers highly sophisticated in this task. For a researcher to be able to attend to the context, the participants and their roles, the timing and structure of observed events, the connections between individuals and their problems, and the individuals' interpretations of their own situations and at the same time coding what the participants fail to say, what they gloss over, and what they ignore and then interpreting imagery making and feelings requires superhuman coding skills.

Collected data are assumed to be gathered by individuals with highly developed and comprehensive observation skills, and, to the extent that such assumptions are not met, the resultant data are invalid. Furthermore, that such observers are able to discern what the events *mean* to the participants, as Charmaz asserts, implies some sophisticated skills in interpretation as well.

Both simultaneous with and subsequent to focused coding is the process called *memo writing*. Charmaz (1983) details this process.

> Memos are written elaborations of ideas about the data and the coded categories. Memos represent the development of codes from which they are derived. An intermediate step between coding and writing the first draft of the analysis, memo writing then connects the bare bones analytic framework that coding provides with the polished ideas in the finished draft. By making memos systematically while coding, the researcher fills out and builds the categories. Thus, the researcher constructs the form and substance toward a finished piece of work and develops the depth and scope of the materials. . . . Sorting and integrating memos follows memo writing. These two steps may themselves spark new ideas which, in turn, lead to more memos. (pp. 120–21)

Sorting leads to memo writing. The integration of the memos follows from their sorting because, as must be repeated, the purpose of grounded theory is to develop theory. Integrating the memos helps develop the relationship (i.e., helps develop the theory). A problem can evolve. Because the researcher starts with nothing—no theory, no hypothesis—he or she has no limitations or operational definitions of variables to determine what data should be collected. A researcher, then, may find as the theory "emerges" that he or she needs more data or needs to define additional variables. This additional data collection is called *theoretical sampling* because it is "sampling [of data] aimed toward the development of the emerging theory" (Charmaz, 1983, p. 124). Its development comes from the inductive process and is evoked by coding and memo writing. The need for it means that the codes (conceptually) and relationships described (memos) have become sufficiently developed that the researcher can examine them in more depth. Theoretical sampling is a way to check those categories and relationships.

The weaving and sifting of categories of variables to formulate the relationships among them allows for, at least, a claim of subjectivity on the part of the researcher and, at most, a gross misinterpretation of facts. The grounded theorists accuse the empiricists of imposing a priori rating-scale values and codes on the participants' responses, while it may be that the grounded theorists' own processual analysis is even more firmly based on researcher bias.

Case-Study Methods

The *case-study method* is one more design strategy under the qualitative rubric. Case studies can be single-participant designs or based on a single program, unit, or school. Merriam (1988) describes how to do case-study research, beginning with translating the research question into more specific and researchable problems, followed by techniques and examples of how to collect, organize, and report case-study data. In addition, she argues that case study is a helpful procedure when one is interested in such things as diagnosing learning problems, undertaking teaching evaluations, or evaluating policy.

Consistent with assumptions of qualitative-research philosophy, the critical emphasis in case studies is revealing the meaning of phenomena for the participants. Stake (1981) acknowledges this assumption, claiming that case-study knowledge is concrete, contextual, and interpreted through the reader's experience. He prefers case-study methods because of their epistemological similarity to a reader's experience. He particularly notes the reasonableness of assuming the natural appeal of the case approach.

Case-study data come from strategies of information collection described in chapter 2 in figure 2 of the qualitative-quantitative interactive continuum: interviews, observations, documents, and historical records. Patton (1980) describes three steps in conducting a case study: assemble raw case data, construct the case record, and write the case narrative (p. 304).

Stake, a well-known advocate of naturalistic inquiry, considers validity to be an advantage of case studies because of their compatibility with reader understanding; in other words, they seem natural. The validity limitations on observation data, previously put forth in the current volume, apply to case studies as well. However, the counterbalancing of information from documents with data from observation and interviews strengthens the resulting validity; invalidity of one set of data can kept in check by considering conflicting or supporting results from the other sources, which is a type of triangulation (a concept discussed later in this chapter).

Enhancing Validity of Case-Study Methods

Case-study methods have potential for increased validity for several reasons. First, because multiple data-collection techniques are used (e.g., interview, document study, observation, and quantitative statistical analysis), the weaknesses of each can be counterbalanced by the strengths of the others, a methodological triangulation. Conclusions related to a certain aspect of a phenomenon under study need not be based solely on one data source. Second, validity may be increased by checking the interpretation of information with experts. Third, case studies generally have a variety of data sources. There should be a structural relationship among these sources. A nomological network should evolve from the empirical materials the researcher gathers. For instance, suppose the demographic data for a school show dramatic changes in the racial profile of the school during the previous five years. The analysis of interview transcripts should be consistent with those data and enlighten the researcher's understanding of those dynamics. To the extent that these findings are consistent within the case, the validity is enhanced. Conceptually, this is similar to giving a battery of tests to obtain an estimate of consistency in the underlying constructs. Fourth, using scientific method, in which a researcher hypothesizes something about the case and collects data to determine if the hypothesis should be rejected, could add to validity and also help future researchers determine starting places for their research. Similar to theoretical sampling, this foreshadowing from data leads the researcher to new understandings. All of these approaches would tend to improve understanding of the case and give in-depth descriptive information.

Interviewing Methods

Patton (1990) characterizes the research interview as a strategy to elicit meaning from the informants that a researcher cannot directly observe.

Interviews can be structured (standardized) or unstructured (nonstandardized), and different interview strategies are warranted by different research purposes (Fontana & Frey, 2005). Newman and McNeil (1998) include a third type, the partially structured interview. The structured interview is designed to collect the same data from each respondent, while the unstructured interview may be used to explore broader issues. In this latter case, each respondent may contribute a different perspective, depending on his or her position regarding the phenomena under study. Unstructured interviews "are totally dependent on the skill and training of the interviewer" (Newman & McNeil, 1998, p. 13). To the extent that such skills are evident, the data collected are likely to be valid. A person who is a keen observer and patient listener is likely to elicit much more information from an informant than an unfocused or hurried questioner.

Structured interviews and partially structured interviews can be subjected to validity checks similar to those used in evaluating questionnaires. That is, are the questions consistent with the purpose of the study? The interview schedule (list of questions) or interview guide is created to direct the interview on a path consistent with the purpose. Diversity of opinion exists about the leeway a researcher may use with the interview guide.

Patton (1980) feels that the guide merely provides

> topics or subject areas within which the interviewer is free to explore, probe, and ask questions that will elucidate and illuminate that particular subject. Thus, the interviewer remains free to build a conversation within a particular subject area, to work questions spontaneously, and to establish a conversational style—but with the focus on a particular subject that has been predetermined. Interview guides can be developed in more or less detail depending on the extent to which the researcher is able to specify important issues in advance and the extent to which it is felt that a particular sequence of questions is important to ask . . . deciding how best to use the limited time available in an interview situation. (pp. 200–201)

Other researchers restrict the guide to a list of questions with a less freewheeling attitude. Hakim (1987) maintains that the validity of this strategy is quite good. As research strategies, interviews provide both more complete and more accurate information than other techniques.

Schatzman and Strauss (1973) consider the interview method valid because they assume that all conversation between the researcher and others at the site is a form of interviewing and that this "naturalness" lends validity to the information obtained. Spradley (1979) might contend that validity cannot be assumed but rests on the quality of the interviewing process the researcher employs. His work, *The Ethnographic Interview,* is the most often cited source for planning interview strategies.[1]

Through probes, follow-up questions, and attention to nonverbal cues, the researcher is able to enhance the data collected. The data are valid to the extent the researcher is able effectively to execute these tasks. Limitations to validity exist, as with other qualitative methods, when the subjective bias of the interviewer affects the interpretation of the data in ways that misrepresent the participants' reality. These invalidations may be more likely with the unstructured interview than with the structured one.

Mouly (1970) points to interviewer bias as the "major weakness" of the method. And so, while some claim that the ability to depart from a rigidly structure probed and follow-up *increases* validity, Mouly would differ.

> To the extent that the interviewer is allowed to vary his approach to fit the occasion, he is likely not only to complicate the interpretation of his results but, even more serious, to project his own personality into the situation, and, thus, influence the responses he received. (pp. 266–67)

Enhancing Validity of Interview Methods

Interviewing can be made a more valid technique in several ways. First, the researcher needs to appreciate that different research purposes warrant different kinds of interviews (Fontana & Frey, 2005). Second, when structured interviews are used, the questions can be checked against the objectives of the study. Third, a high level of interviewer training increases validity. Fourth, when several interviewers participate, a variety of interpretive perspectives may add depth to the interpretations. Fifth, checking for consistency across participants increases reliability, which adds to validity. Sixth, debriefing the interviewers and informants after data collection can help increase validity. From the perspective of Fontana and Frey (2005), the interview is "negotiated text" (p. 716). In the negotiation process, the researcher reviews dialogues with the informants to check the interviewers' own biases, because interviews are not neutral, objective, and detached from the interviewers' own perspectives, beliefs, and values. Current trends in interviewing appreciate the interview as co-constructed between interviewer and informant (Fontana & Frey, 2005).

One definition of validity has to do with its ability to predict and explain underlying constructs. Once interview data are collected, a researcher can determine how well the interview explains certain underlying constructs related to the purpose of the interview. If the researcher has hypotheses or assumptions to start with, the data can be used to see if these assumptions are verified (predicted) or contradicted. Based on these new findings, either the theory is supported and new assumptions are formed, or new directions for future research are suggested, or both. As is apparent from this volume's discussion, the validity of qualitative methods frequently is increased by using what are considered traditional quantitative methods.

On the other hand, Kvale (1996) distinguishes interview transcripts from "data." The two, he argues, are not similar. Transcripts are "abstractions," which are constructed from the original interaction between the interviewer and informant. Those narratives cannot be considered objective data. He offers a model of seven stages of the interview process. Fink (2000) also suggests that in addition to the transcript, the recall of the interviewer is important. Field notes supplement the meaning of the transcripts. As the researcher codes the transcript with interpretive tags and labels, his or her recall is aided by these field notes. Audiotaping interviews is a common practice, and videotaping captures even more meaning (Fink, 2000).

Historical Methods

Historical research is carried out primarily through document study and interview techniques. There is disagreement about the scientific status of historical research methods, and several of these points of view are reviewed in this section. Some maintain that at all times, the historian operates inductively, drawing from data to formulate conclusions as the research is carried out. No hypothesis is initially stated, and historians frequently have to reconstruct the facts from unverifiable sources, according to Mouly (1970). These facts are based on their plausibility and can only be inferred; they cannot be measured. Furthermore, because the design necessarily focuses on unique events, a researcher is unable to generalize. Shafer (1974), in an interesting treatise detailed later in this section, maintains that there are facts that we do accept from a historical perspective, and he discusses the historian's search for causation and the place of statistical analysis.

Good (1963) asserts,

> The historian thinks of the method of investigation as scientific, and the manner of presentation as belonging to the realm of art. These interrelationships of history and science make it important that the modern historian be well grounded in the natural sciences. (p. 181)

One would expect that historians of Good's description would consider and deal with threats to validity. He assumes, differently from Mouly, that historiography is a science: "History qualifies as science in the sense that its methods of inquiry are critical and objective, and that the results are accepted as organized knowledge by a consensus of trained investigators" (p. 181).

As Good describes the process of historical research, it does not differ substantially from the processes of quantitative research. However, he says, historians and quantitative researchers start at different places. Historians do not do direct observations or experimentation, but they do utilize "reports of observation that cannot be repeated." Good describes that process.

> Historians cannot recall the actors of the past to reproduce the famous scenes of history on the stage of today.... Therefore, the historical method involves a procedure supplementary to observation, a process by which the historian seeks to test the truthfulness of the reports of observations made by others. (p. 183)

Even more closely aligning the historians with quantitative researchers and with the traditional scientific method, Good compares the two and then describes the uniqueness of the historian.

> Both historian and scientist examine data, formulate hypotheses, and test the hypotheses against the evidence until acceptable conclusions are reached. A number of historians, in emphasizing

the interpretation and meaning of facts, have sought to identify tendencies, themes, patterns, and laws of history, while some of these investigators have dealt with such philosophical or theoretical problems in history as discovery of laws, unity and continuity, possibility or impossibility of prediction, and oversimplification growing out of the search for clues or keys. (p. 183)

In his discussion of historical strategies, Shafer (1974) acknowledges the bridge over the qualitative-quantitative gap made by the historian's use of data-analysis methods. He uses as his example the African American vote in Virginia for Lyndon B. Johnson in the presidential election of 1964. Within this historical study of that election, a researcher might measure the vote count for Johnson and for Barry M. Goldwater in districts with high and low rates of African American registered voters. From these data, the researcher could test a hypothesized relationship in Virginia's polling districts between the rate of African American voter registration and the pro-Johnson vote. Thus, what would traditionally be considered a quantitative technique appropriately fits this research and adds validity to the study, demonstrating another advantage to denying the methodological purity of either end of the qualitative-quantitative continuum.

Mouly's assertion of nonverifiability of historical facts is tempered somewhat if the investigator has access to primary sources (i.e., original and factual documents). Invalidity originates from the secondhand nature of data; and, to the extent that original sources are available, the invalidity is reduced. Also, checking facts when possible with multiple sources would reduce the invalidity of this method.

Many processes (e.g., Mouly, 1970; Shafer, 1974; Wiersma, 1980) of external criticism and internal criticism are steps in the historical researcher's work. Briefly stated, while these processes overlap, *external criticism* (called a *validity check* by Wiersma) deals with the genuineness or authenticity of the document used. This is a step prior to *internal criticism* (i.e., the determination of the meaning and trustworthiness of statements within the document). The distinction is necessary because, although an original source is authentic, statements within it may not be completely accurate; they may be overlaid with the author's bias or political stance. Because document study is a predominant aspect of historical methods, these steps are critical.

Shafer adds a third step, *synthesis*, the blending of evidence resulting from the first two steps to report historical events with accuracy. Shafer admits that while external criticism and internal criticism are validity processes, this third step is "necessarily riddled with subjectivism" (1974, p. 26). One must hope that it is not riddled enough to invalidate the outcomes of steps one and two.

Shafer (1974) acknowledges two stances regarding the historian's place on the qualitative-quantitative continuum: Mouly's perspective (history as a nonscientific mode) and Good's perspective (history as a scientific mode). He describes these two points of view:

> One is that the historian's cultural experience or environment affects his interpretation of evidence on human affairs, and that as a result interpretations of history necessarily vary with the social environments of historians. The other is that inference (i.e., supplying data not explicitly provided by the contemporary writers or artifacts) must be indulged in with great care. (pp. 12–13)

In other words, the validity of historical methods is based on two assumptions: that the interpretation of history varies with the subjective social experience of the historian and that the reporting of history should not go beyond the database. Some moderation of this relativistic dichotomy of historical research occurred in the mid-1900s, according to Shafer, when, in essence, it was recognized that there are factual aspects of societal life that we can know and know with confidence. Therefore, the work of historians need not always be questioned (by the scientifically based) on validity grounds. Indeed, this epistemological conclusion assigned a new level of validity to historical findings that quantitative researchers could accept. As Shafer (1974) explains it,

> On the theoretical level, relativism in the middle years of the twentieth century was rather modified by arguments by epistemologists and others that human activity shows some "probabilistic" regularities, permitting assumptions and explanations in which we may repose considerable confidence. (p. 13)

Some "purist" quantitative researchers might read that conclusion and see it as the historian's affirmation that human behavior, as well as social phenomena, is lawful, measurable, and based on a scientifically discoverable theory that can be revealed by controlled designs for hypothesis testing. And this is exactly what the quantitative researchers have been trying to convince the qualitative researchers of all along. In any event, this conclusion by some—that historical research now subsumes the quantitative philosophy—should not be as a rationale to be negligent in estimating validity when utilizing historical methods.

Shafer goes to considerable lengths to detail the threats to the validity of historical research and the nuances of analysis to which one must attend. Three cautions in particular are here. First, while historians struggle to delineate causes of events of the past, causation can never be concluded. Historians must examine immediate precipitating events, as well as underlying events, may be related to the historical phenomena under study. Second, relations may be found between precipitating events and the events being studied. Third, he acknowledges that multiple causation is more relevant to history than unidimensional causation. Note that these three tenets, explained at great length by Shafer (1974), also are three of the basic tenets of quantitative research: (1) causation can rarely be proven; (2) *correlational* is not synonymous with *causal;* and (3) multivariate relationships are more common than univariate relationships.

In summary, historical research not only encompasses methods already described (interview, case study) but also utilizes document study as a major strategy. Historians must seriously consider validity and its threats, as must all other researchers. Utilization of data-analysis techniques in a traditionally quantitative sense can be useful in increasing validity.

Enhancing Validity of Historical Methods

One of the major problems with historical research is its inherent validity weakness, inherent because events have occurred in the past outside of the possible experiences of the researcher or the participants, and so verification is always of a secondhand nature. Historical study can be made more valid by including checks on sources of data and ensuring that multiple sources are used.

Ödman, a professor at the University of Stockholm, studied witch trials in Stockholm, Sweden, during the 1670s. The reader is directed to his example of applying his model of interpreting historical evidence in an issue of *Qualitative Studies in Education*. He discusses the validity problem in historical interpretation, grounding his ideas in the work of Trankell (1972) and Ricoeur (1988). Ödman's (1992) discussion delves into the situations in which actual physical reality exists, as well as situations in which the researcher works at a symbolic level. Ödman's model should be reviewed by those pursuing historical research in education. The reference here to his work is insufficient for understanding it in depth.

To the issue of validity, Ödman calls on his predecessors for criteria.

How the validity problem is approached in an individual case depends partly on which kind of reality the interpretation is concerned with. . . . Two criteria which Arne Trankell formulates in his *Reliability of Evidence* (1972) are useful. . . . The first one can be summarized: *If an interpretation leaves an essential part of the information unexplained, this interpretation cannot be accepted as a valid description of the reality which the data are referring to . . . the second formal principle: If an interpretation is to be accepted as a valid description of the reality the data are referring to, it must be the only one that gives a complete and reasonable explanation of the information available.* [italics in the original] (p. 170)

Ethnographic Methods

Ethnography is literally a "written account of a culture." It serves as a strategy for studying the commonsense features of everyday situations—the common happenings in a particular setting of interest. In these studies, interaction as an ongoing process is scrutinized and recorded in descriptive detail (Creswell, 2005; Fetterman, 1989; Trueba 1991; Trueba, Jacobs, & Kirton, 1990). Creswell (2005) claims that ethnographies usually include cultural themes; descriptions of shared patterns of behavior, belief, language, themes, and interpretations about the context or setting; and researcher reflexivity (p. 442).

Ethnography, whose roots are in anthropology, has a forty-year-plus history of being seriously utilized in adapted ways in education. In the early 1960s, Stanley Diamond at the New School for Social Research in New York City established a series of conferences on anthropology and education. This program, called the Culture of Schools Program (Wax, Diamond, & Gearing, 1971), was funded by the U.S. Office of Education (USOE). In 1965, the program was moved to the sponsorship of the Anthropological Association, with USOE support intact.

This, then, evolved into the Program in Anthropology and Education (PAE), subsequently directed by Fred Gearing, who, with M. L. Wax, developed ongoing conferences. In 1970, the Council on Anthropology and Education formally organized and began a newsletter.

While the cultural aspects of educational activities are now investigated through anthropologically based methods, the traditional training one receives in education research is not comparable to the six-year training an anthropological researcher typically receives. A research training seminar entitled Anthropology Field Methods in the Study of Education was offered for the first time at the American Educational Research Association's annual meeting in 1968 (Sindell, 1969).

The data-collection techniques within ethnography consist primarily of participant observation, along with the strategies used in case studies and grounded theory. Spindler (1974) calls participant-observer methodology "the keystone of an anthropological approach" (p. 385). Therefore, the limitations on validity are those that have been discussed for those methods.

Many would maintain that validity is the major strength of these methods. LeCompte and Goetz (1982), in particular, point out four reasons for the high internal validity of these strategies, although external validity, the degree of generalizability, is neglected. Internal validity comes from the following methodological features:

- Ethnographers commonly act as participant observers and live among the participants over long periods of time; they are able to continually refine their interpretations over time and compare them to "reality."
- Interviewing informants involves using phrasing and vocabulary more closely in tune with the participants' own and also less abstractly than instruments used in quantitative studies. This, therefore, increases the likelihood of the instrument being able to tap the information for which it was developed.
- Participant observation is conducted in natural settings that are the reality of the life experiences of participants more so than are contrived settings of quantitative studies.
- The analysis in ethnography uses a process of "researcher self-monitoring," a "disciplined subjectivity" that brings the study under continual questioning. (p. 43)

Three reasons that ethnographers ignore external validity, according to LeCompte and Goetz (1982) and Goetz and LeCompte (1984), are, first, the purpose is to describe in detail aspects of a single participant, group, or unit. Even if multiple sites are used, the researcher is obligated to enter each site as if he or she had no other information and as if this site were unique. Generalizability, then, is precluded. Second, the ethnographer enters the field site without assumptions, preconceived notions, or hypotheses. Therefore, there are no bases for comparison or generalizability. Third, the problem studied, the nature of the goals, and the application of the findings differ substantially from traditional quantitative methods, and so definitions of external validity must vary.

Concerning the problem studied, the credibility of quantitative designs is based on examining effects in *controlled* situations, looking at variables uniquely, one at a time. In contrast, according to the discussion in LeCompte and Goetz (1982), ethnographers focus on the "interplay of variables in natural context." Credibility is based on examining "all causal and consequential factors" (p. 33). Regarding the goals of studies, the goal of ethnographic research is to develop theory not to test it, which requires that a priori relationships be avoided. While quantitative researchers aim to generalize from the sample to the population, and external validity must (by definition) be built in, ethnographers aim for comparability and translatability in order to apply their results. Comparability differs from generalizability and "requires that the ethnographer delineate the characteristics of the group studied or constructs generated so clearly that they can serve as a basis for comparison with other like and unlike groups" (p. 34). Translatability is the aim—the quality that "assumes that research methods, analytic categories, and characteristics of the phenomena and groups are identified so explicitly that comparisons can be conducted confidently" (p. 34).

Phenomenological Research

Not too different from the methods described thus far is phenomenological research. As described by Lincoln, on the premise that one socially constructs reality, the phenomenologist "looks in natural contexts for the ways in which individuals groups make sense of their worlds" (1990, "Toward," p. 290). The logical researcher "collects" the realities of the participants and the interpretations of their constructions. Lincoln distinguishes this researcher's task from the quantitative researcher's by its expansionist purpose; it is contrary to the numerical reductionism of quantitative researcher's work. Van Manen (1990) describes this as "human science."

> From a phenomenological point of view, to do research is always to question the way we experience the world, to want to know the world in which we live as human beings. And since to know the world is profoundly to be in the world in a certain way, the act of researching—questioning—theorizing is the intentional act of attaching ourselves to the world, to become more fully part of it, or better, to become the world. (p. 5)

Kvale (1983) describes three aspects of this method: open description, investigation of essences, and phenomenological reduction. The step *open description* is for the participant merely to describe his or her experience as completely as possible, extemporaneously, with no consideration of cause or origin. The second step, *investigation of essences*, involves, as Kvale puts it, "varying a given phenomena [sic] freely in its possible forms, and that which remains constant through the different variations is the essence of the phenomenon" (p. 184). Finally, *phenomenological reduction* involves "suspension of judgment as to the existence or non-existence of the content of the experience" (p. 184). Simply stated, in this particular phase of the method (sometimes called *bracketing*), the researcher puts in parentheses his or her foreknowledge and common sense about the emerging phenomenon to help arrive at the essence of it. To relate it to quantitative-research philosophy, this operation is a way of accounting for the researcher's prior knowledge, subjective biases, and expectations. Kvale puts it this way: "The phenomenological reduction does not involve an absence of presuppositions, but a consciousness of one's own presuppositions" (1983, p. 185).

Howard (1994) uses a phenomenological-research approach to study the meaning of adults confronting computer technology for the first time.

> This, then, is a phenomenological investigation of the experiences of those persons who have sat down to the keyboard and can articulate the experience of being introduced to a computer. Another

interpretation of the "text" may help us "see" computer technology as if for the first time. (p. 34)

Even though phenomenology is private and holistic, it can be defended as *not* antiscience. Because human experience is unique, one cannot detach and reduce external data; further, Kvale (1983) implies, it is this inability to abstract that forms the existential nature of psychology. That is, general laws and theories cannot be applied to a specific individual in a unique set of circumstances. Others, however, argue that emerging sets of themes from many participants may form the essence of a generalization applicable to those in similar states of life.

In an in-depth phenomenological study of educational leadership, Mitchell (1990) explains that he selected this approach to inquiry to get at the "lived experience" of educational leaders. Based on the original ideas of Edmund Husserl over eighty-five years ago, Mitchell assumes that philosophy can be independent of objective science and that "the ultimate foundation of all knowledge is in human consciousness" (p. 255). Because phenomenology is not like the scientific method, a unified system or set of methods that can be learned, it is "more a method of study, a way of viewing, a perspective, a stance, a manner of thinking ... and its basic tool is 'seeing' and 'interpreting' what is seen" (p. 253). Mitchell cautions that researchers who apply the phenomenological approach in education are not all engaged in the same pursuits.

Appendix A is a summary of a study on humor by Foerstner, Newman, and Koenig (1985) that demonstrates one method of phenomenological research. In this example, the participants are allowed to discuss humor freely and in their own words, while they alone structure their responses. In the analysis, the transcription is carefully reviewed while attempting to maintain maximum openness. Following the first reading of the responses, the central meaning expressed by the participants is explicated. Then they are related the whole to get at their central themes, their essence.

Enhancing Validity of Phenomenological Methods

Wertz (1986) makes the claim that for reliability in logical research (the process of defining essential themes from informant experiences), the themes must be present in every informant's experience. They are "invariant despite changes in the factual details" (p. 197).

One method of enhancing validity that crosses many other methods is *triangulation*. Triangulation is the combination of several data-collection strategies or data sources in the same design. Jick (1979) traces the concept of triangulation in social science to what Campbell and Fiske (1959) call *multimethod-multitrait research strategies*. Its use in the social sciences stems from the concept of triangulation in the military. Simply put, in navigation strategy, using multiple reference points (and not necessarily three) enables one more easily to pinpoint an object's exact position (Denzin, 1978).

Jick (1979) describes the *between-methods* type and the *within-methods* type of triangulation. The between-methods type is

> a vehicle for cross validation when two or more distinct methods are found to be congruent and yield comparable data. For organizational researchers, this would involve the use of multiple methods to examine the same dimension of a research problem. For example, the effectiveness of a leader may be studied by interviewing the leader, observing his or her behavior, and evaluating performance records. (p. 602)

Between-methods triangulation is the most conventional form and is used to test the data for the degree of external validity. Its frequent use is based on this fundamental assumption: "The effectiveness of triangulation rests on the premise that the weaknesses in each single method will be compensated for by the counter-balancing strengths of another" (p. 604). And, for all practical purposes, this type can be assumed to be the standard usage of triangulation. The second type, within-methods triangulation, is rare. It deals with reliability and "essentially involves cross-checking for internal consistency" (p. 603). This form of triangulation is much less frequently used because it is limited to the use of just one method.

Essentially, a researcher would use several variations of one method to collect several sets of data, which would then be compared. Jick (1979) reviews the advantages and disadvantages of triangulation. First, the advantages:

- allows researchers "to be more confident of results" (p. 608)
- can stimulate creative methods, new ways to "capture" a problem (p. 608)

- can help "uncover the deviant or off-quadrant dimension of a phenomenon" (p. 609)
- can lead to "enriched explanations" of research problems (p. 609)
- can lead to a "synthesis or integration of theories" (p. 609)
- can serve as a test of competing theories (because of its comprehensiveness) (p. 609)

Second, the disadvantages:

- replication very difficult, if not impossible (p. 609)
- is of no use if the wrong question is being asked; "If the research is not clearly focused theoretically and conceptually, all the methods in the world will not produce a satisfactory outcome" (p. 609)
- should not be used "to legitimate a dominant, personally preferred method. . . . If either quantitative or qualitative method becomes mere window dressing for each other, then the design is inadequate or biased" (p. 609)
- must justify the use of the multiple methods (e.g., cannot assume all are equally sensitive to the phenomenon being measured) (p. 609)
- "may not be suitable for all research purposes" (p. 610)
- "demands creativity" (p. 610)
- is expensive in terms of both time and cost (p. 610)

Using multiple sites is one technique suggested by LeCompte and Goetz (1982) to increase validity. Admittedly, though, the examples they provide do not meet the standard for generalizability that quantitative assumptions would require. The extent to which four factors—*selection effects, setting effects, history effects,* and *construct effects*—are present reflects the increased validity of the study, according to LeCompte and Goetz. *Selection effects* simply force the ethnographer to compare only constructs among groups in which they occur. The first issue for the researcher is to match the phenomena under study with the nature of the groups. If not carefully done, a researcher might begin a study under inaccurate assumptions regarding the nature of the groups at the various sites. *Setting effects,* too, can be diminished. An example given by the writers is that using teachers as observers in classrooms resulted in a teacher-classroom interaction that made their data characteristically different from the classroom-observation data of nonteachers. To help counteract this effect, LeCompte and Goetz suggest that both a teacher and a nonteacher observe and report. Setting effects also occur when groups are frequently under study (such as happens when schools are near universities), which can be counteracted by choosing nonresearched groups. *History effects,* when counteracted, will increase validity. When using more than one site in ethnography, the various historical foundations of those sites need to be acknowledged. Three nursery schools may be studied in detail using ethnographic methods, but the development of each may have evolved from extremely different roots. *Construct effects* occur under several conditions. First, when constructs under study are idiosyncratic to groups under study, appropriate comparisons to the other groups diminish, that is, the ability to make appropriate comparisons to other groups is lessened. Second, to the extent that the use of any observational instruments is not common across groups, there are likely to be construct effects. Third, the meanings of phenomena might vary across groups, creating construct effects.

By attending to the methods to improve trustworthiness in chapter 3, one is more likely to become sensitive to the design validity of a study. The more evidence of design validity a study can show, the more truth value we can presume. Similarly, these concerns are revealed in quantitative studies as concepts of internal and external validity (Campbell & Stanley, 1963). With a predominantly quantitative study, the stronger internal and external validity a study has, the stronger the truth value of the study.

A good researcher needs to be familiar with a variety of methods. Multiple methods may enhance the quality of a research study. Lombard (1991) demonstrates how the use of multiple methods increased the quality of her research. In her case, however, all of the methods she uses tend to be qualitative: interviews, member checks, and critical-incident techniques. Stivers and Srinivasan (1991) also use multiple methods to improve their insights and the interpretability of research, and they use mathematical algorithms and mathematical simulations to enhance and support their interpretations. Increasingly, researchers are using multiple methods (mixed methods) to improve the quality of their research. Doing so is consistent with the qualitative-quantitative continuum, which is based upon the assumption that investigators should not be tied to any single methodology. The research purpose and research question always should dictate the method.

Applying the Qualitative-Quantitative Interactive Continuum

This chapter begins with procedures that can be used in critiquing research, with actual examples of research published in the disciplines of education and counseling, contextualizing them within the continuum so that their validity can be evaluated.

The process of critiquing each study involves assessing the methods the researchers use; the methods are presented in figure 3. The totality of the "methods" we call the *design*. When critiquing a published study, one is limited to knowing only what is written in the article about the methods the researcher uses. Full accounting for each activity on the part of the researcher may or may not be included. Our judgments about each of these studies are limited, therefore, as are all critiques of published work.

More important than the conclusions we draw about these four studies is the process we suggest. Others may ask somewhat different questions about each study. We are not particularly bothered by that. Our questions here are not uniform from study to study. The bottom line for us is advocating for a critique of published research that seeks to judge whether the research question is consistent with the research methods. Our process is only one of several that can accomplish that goal.

Chapter 5 includes

- discussion of why utilizing the continuum increases the researcher's awareness that research is a holistic endeavor
- discussion of posing and answering consistency questions across a research study in the areas of purpose and methods, methods and data, purpose and conclusions, and implications and purpose
- application of the continuum to a published research study
- application of the continuum in planning a research study

Procedures to Use in Critiquing Research

In evaluating research studies, the researcher can apply the continuum as an assessment scheme. In planning a research study, the researcher can utilize the continuum to assess his or her plans. And, because research is conceptualized as an unbroken continuum, one may enter the continuum and make inquiries for assessment and critiquing purposes at any point. The following steps are a place to begin, especially when one is initiating the research process. In the *consistency-questions model,* one asks,

- What is the question, purpose, or reason for doing the research?
- What research methods might one use to address this question, purpose, or reason?
- Contingent on the answer to the second question, what are the underlying assumptions of the research method?
- What are the findings of the research?
- What are the implications of the findings of the research?

As illustrated in figure 4, the sequence of questions is linked, implying continuity and consistency. Because of this mapping, a researcher is able to assess the consistency from any point in the loop to the adjacent point: Is there a match between the question or purpose (A) and the methods (B)? Is there a match between the methods (B) and the assumptions (C)? Is there a match between the implications (E) and the purpose of the question (A)?

For example, one might carry out research, as in the early-school-effectiveness studies (Olson, 1986), in which the purpose is to explore what characterizes a school where learning gains are relatively high. To acquire thick descriptive detail, this question dictates the use of qualitative methodology. Following such a study, however, the investigator should not generalize from the descriptions to other schools. Generalization is consistent neither with the purpose of the study nor with the underlying assumptions of the specific research methods. If one were to generalize, the generalization would be criticized as inappropriate and in violation of the assumptions.

This consistency-question approach as depicted in figure 4 is a subset of the interactive continuum represented in figures 1 and 2. The continuum assumes that the research question dictates the methodology.

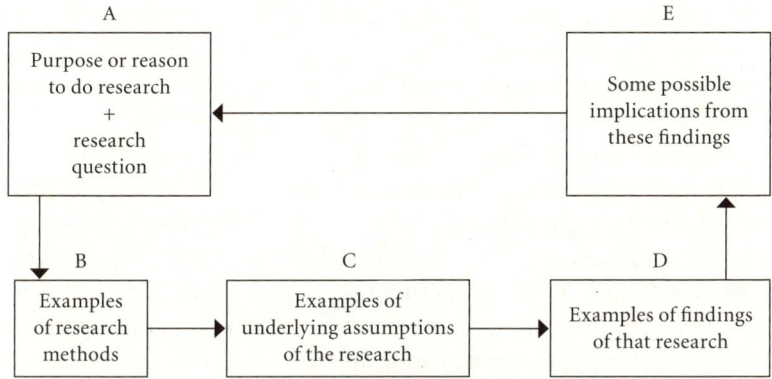

Figure 4. Model of the consistency questions to ask in critiquing research. Do the research question and research purpose dictate the research method?

If the researcher uses methods consistent with his or her purpose, the conclusions likely will be consistent with the underlying assumptions of those methods. The consistency questions to ask are common to all research studies.

Figures 5 and 6 illustrate possible answers to the consistency questions when either the qualitative or the quantitative paradigm is preeminent. The two figures would ideally be depicted in one schematic drawing, which would be more in keeping with the continuum. They have been separated here for illustrative purposes. For example, if a researcher asked question A in figure 5 or 6 ("What is the research question?"), the answer (that the researcher wanted to test a set of hypothesized relationships) would lead to the point on the continuum illustrated in figure 1 corresponding to that purpose, square 3. Examining figure 1 at square 3 shows that testing hypotheses is derived from a review of the literature, square 2, and is followed by collecting the data, square 4. The researcher who has defined his or her purpose as describing a certain phenomenon in detail, with no preconceived hunches or hypotheses, would enter the research continuum in figure 1 at circle A. As is apparent in figure 1, this is the first step in utilizing the qualitative part of the continuum. It is followed by analyzing the data (circle B), drawing conclusions (circle C), attempting to derive hypotheses (circle D), and perhaps developing a theory (circle E) that places the research within the area of overlap with the quantitative part of the continuum. Using the continuum forces the investigator to perceive the research in a holistic context, in a context of both qualitative and quantitative assumptions rather than within a narrow bias of either one or the other.

The consistency questions (figure 4) assist the researcher in ensuring that consistency exists throughout the research, from questions or purpose through implications. In figures 5 and 6, the questions are placed so that consistency is maintained throughout both paradigms (along the continuum). Accepting the continuum implies accepting that consistent with the central place of theory, all other components must coexist in appropriate relationship to it. Adopting the model and planning research within its structure permit research to be carried out consistently and, thus, with optimal validity. Researchers who adopt the model assume a philosophy of research that is unified and focuses on question and method consistency, not on ideology.

To conceptualize the logic in applying the interactive continuum, we suggest three phases that flow naturally from what has just been presented. In the first phase, the continuum assumes consistency between question and method. In the second phase, one evaluates the extent to which there is consistency. And in the third phase, each study can be examined closely for issues of design validity.

The next four sections present suggestions for applying the continuum and its subset of consistency questions and present assessing issues of design validity to evaluate research. Four published studies are critiqued (appendixes B through E). We want to emphasize that the process being suggested can be generalized to virtually any situation. It can be adapted, used flexibly. We emphasize the *process* of critique—asking the questions about the research studies one reads is the important activity. Two people (or three and so on) may come to different judgments about the truth value of any one article. Each critique is a value judgment. The four articles here may be critiqued by our model, but their conclusions might differ in some limited ways.

These critiques exemplify the second and third phases of research evaluation, that is, examining issues of design validity as defined at the end of chapter 3. We use our model to critique each article and then add more general reflections after each critique. The reader will notice that the four critiques, while following our model conceptually,

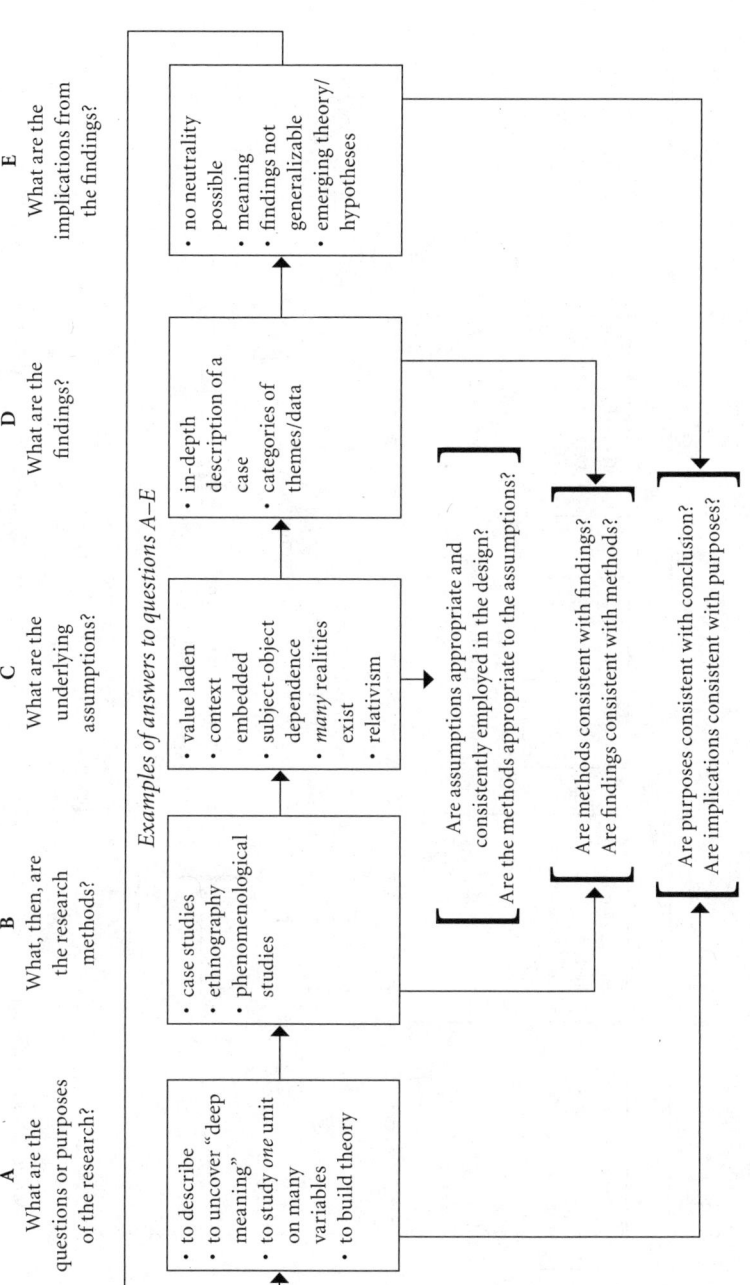

Figure 5. Model of the consistency questions to ask in critiquing research, the qualitative paradigm being dominant

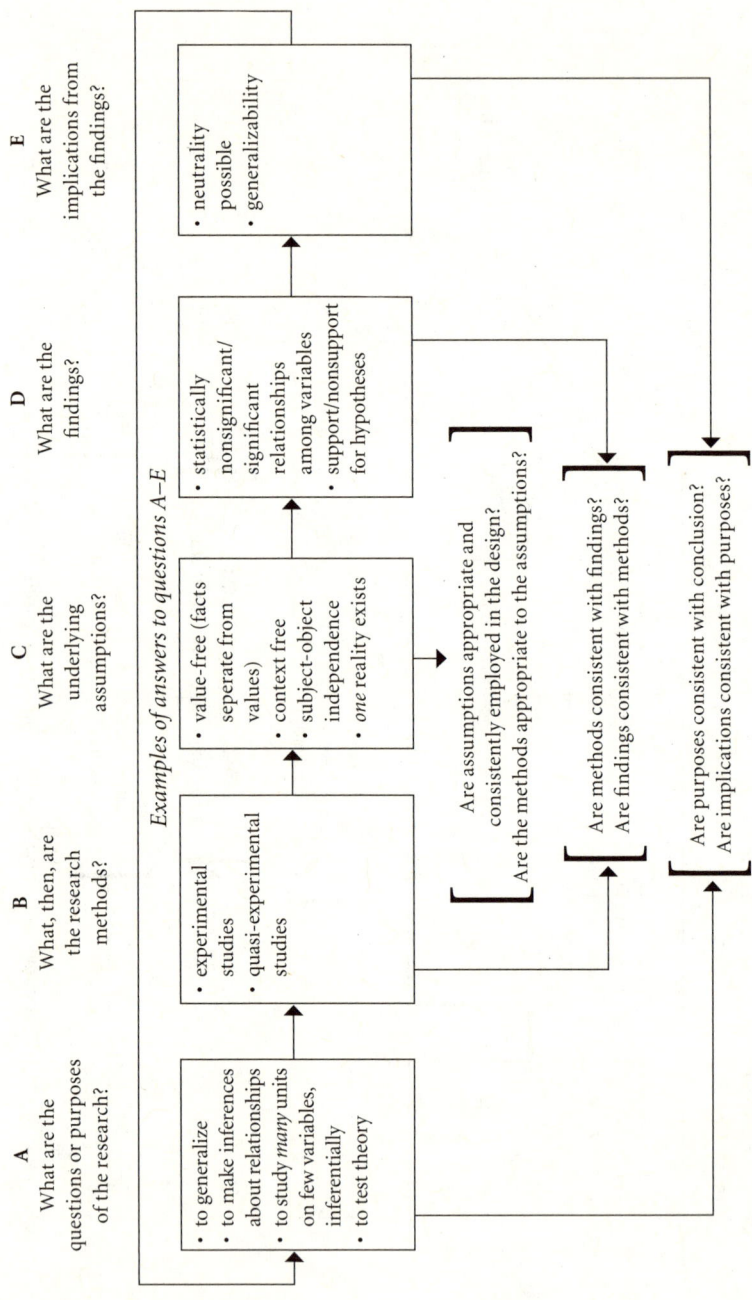

Figure 6. Model of the consistency questions to ask in critiquing research, the quantitative paradigm being dominant

vary widely in how they are carried out. We are less concerned with the specifics than with the value of such critiques and that they are actually conducted.

Critique of the Alexander and Harman Study

The first study to be critiqued, "One Counselor's Intervention in the Aftermath of a Middle School Student's Suicide: A Case Study," by Alexander and Harman (1988), can be found in appendix B. The reader will find the critique more meaningful if the article is read before continuing.

The purpose of this study is to describe an intervention for middle school students who were affected by a classmate's suicide. There is no statistical analysis, and the authors indicate that attempts to generalize should be avoided. This, then, clearly falls into the broad category of qualitative research, and we analyze it in terms of the qualitative-research components previously identified.

Neutrality

In terms of neutrality, this study is weak because there is only one observer, and there was no apparent attempt to control for personal bias.

Prolonged Engagement On-Site

Prolonged engagement on-site seems to be sufficient for this study because the counselor was on-site prior to the suicide and counseled some students for up to twelve weeks following the incident.

Persistent Observation

There is no way to assess the students' behavior before as compared to after the suicide. The inference of this study is that there was an increase in suicide ideation and a fear of additional suicide attempts. No data are presented to substantiate this claim.

Peer Debriefing

According to the study, there was no attempt at peer debriefing. No other counselors were sought to check personal perceptions or bias or to offer other interpretations.

Triangulation

There was no attempt to obtain information from other sources.

Member Checking

As a result of the counselor's interaction with the students, she concluded that certain themes emerged, such as poor self-concept and excessive self-demands. There is no report of checking these themes for persistence in or consistency among group members.

Referential Materials

There is no way to assess whether referential materials were used—none is identified, and we can only assume none could be made available.

Structural Relationships

No other data are shown to have been available to interweave to establish structural relationships between data sets. The counselor does, however, attempt to interweave Gestalt theory into her conceptualizations.

Theoretical Sampling

No attempt at theoretical sampling is described.

Leaving an Audit Trail

There is no indication that notes, records, or any other kind of documentation was kept. It appears to be entirely personal observation.

Generalizability

Applicability. To estimate the applicability of this study, one needs deep descriptors to clearly define the characteristics of the sample. Description is not of sufficient detail to have a clear sense of socioeconomic status, culture, and so on.

Context limited. There is no indication that the interpretation is context free. On the contrary, there is a good possibility that it is context specific. The concepts that the authors apply are basically Gestalt, and the interpretations are all from this perspective, as is the investigator's training and predisposition.

Replicability. No data are available to estimate the consistency of the findings.

Negative-Case Analysis

No evidence is available.

Truth Value

The purpose of this study is to identify the potential of a Gestalt approach in helping children who may be the trauma of a classmate's suicide. In light of the above validity criteria used in evaluating qualitative research, we have limited information and are skeptical about the study as it relates to making a strong statement that Gestalt is a viable intervention approach in these situations.

Reflections

By reflecting on Alexander and Harman (1988) through the framework of the interactive continuum, one can gain insights into both the strengths and the limitations of its design validity, From these insights, future researchers interested in replication, when making decisions about methods, might want to consider what has been revealed.

This article is appealing in that it is well written, intuitive, and deals with some obvious truths about the need to be sensitive to children who have experienced a suicide. However, it is methodologically weak, at least insofar as what is reported in the published article. Reporting more thoroughly on all aspects of methods and improving on those that are omitted would strengthen its potential impact on knowledge about intervention strategies. Future researchers may want to take these considerations into account when designing a study.

Given what we know about the study, there are techniques that could have enhanced the validity of this study. Alexander could have enlisted the aid of other observers, which would have helped counterbalance the effects of personal bias and sensory deficit. Checking her interpretations with experts in the field also could have helped make the study more valid; and the use of different forms of data collection to supplement the unstructured interviews would have been helpful.

If one applies the interactive continuum of qualitative-quantitative methodology, the themes that emerge from this case study are clearly qualitative, yet these themes provide feedback appropriate to the quantitative end of the continuum. The themes could become the basis for a study more empirical in design. Students could be randomly assigned to treatment groups, such as Gestalt, behavioral, and no treatment (control, perhaps a group experience based on a self-help model without formal theoretical basis). The themes that emerge from the case study could then be used to evaluate treatment effectiveness while controlling for such variables as age and sex to increase the ability of the tests to determine any effect due to treatment alone. Alexander and Harman's mix of qualitative and quantitative characteristics is not in itself a problem, but their study might have been enhanced had they elaborated upon this mixture and had they submitted variables developed by the case study to an empirical investigation. A strength of this study is that it sensitizes people to an important topic. The research also may have heuristic value in that it may lead to further study. However, before one could conclude that Gestalt therapy is effective, more control in the research procedures would be necessary

Critique of the Curtis Study

The second study to which we apply the continuum is "Effect of Therapist's Self-Disclosure on Patients' Impressions of Empathy, Competence, and Trust in an Analogue of a Psychotherapeutic Interaction" by John M. Curtis (see appendix C).

The critique of the Curtis (1981) study is a composite by students in a PhD counseling research seminar.[1] While similar to the other critiques, there are some real differences. Different evaluative questions are used, for example. The students analyzed the study by describing the research question, the statistical questions, or hypotheses; the research design in Campbell and Stanley's (1963) terms, the relationship between the research question and the design, the statistical model, the conclusions of the study, the recommendations that the students would suggest, general quantitative considerations, and general qualitative considerations. We include this critique to show that it is the process of evaluating a published research article that is important; the specific format one uses is less important.

Research Question

Is there a relationship between a therapist's self-disclosure and patients' perceptions of the therapist's empathy, competence, and trust?

Hypotheses

1. High therapist self-disclosure condition will yield patients' lowest evaluations of empathy, competence, and trust.
2. Low therapist self-disclosure conditions will yield patients' moderate evaluations of empathy, competence, and trust.
3. No therapist self-disclosure conditions will yield patients' highest evaluations of empathy, competence, and trust.

Research Design

Refer to symbols described in chapter 3.

   R X_1 O X_1 = high disclosure
   R X_2 O X_2 = low disclosure
   R X_3 O X_3 = no disclosure

THREATS TO INTERNAL VALIDITY

Threats from history, maturation, mortality, instrumentation, testing, and statistical regression were controlled for. Selection bias was uncontrolled, as volunteers who were ongoing clients agreed to participate in the study. The author fails to explain randomization procedures.

THREATS TO EXTERNAL VALIDITY

1. Reactive effects
   a. Posttesting—no problem indicated
   b. Subjects aware of participating in research; could have had *reactive effect,* creating bias in responses (Hawthorne effect)
2. Generalization to population: outcomes can only be generalized to people who meet five criteria indicated in study
3. Generalization to other settings: dependent and independent variables *not* sufficiently defined by author; no sample provided of independent variable

CONSTRUCT VALIDITY

1. Author's definition of self-disclosure defined narrowly and not necessarily consistent with other theorists' views of construct
2. Author's operational definition of self-disclosure not consistent with psychodynamic view, which is assumed by investigator
3. Operational definitions of constructs measured by test instruments not defined
4. Two separate measures of empathy used without defining construct of empathy or how each instrument defines empathy

Relationship between Research Question and Design

The results of the study are highly misleading regarding therapist self-disclosure. In effect, the research design does *not* reflect the research question; that is, self-disclosure is not defined accurately.

Statistical Model

The author uses three one-way analysis of variance procedures (ANOVAs) without indicating how he corrected alpha level for multiple tests. Alternatively, he could have used the general linear model multivariate analysis of variance (MANOVA, or factor analysis).

Statistical Answer/Results and Conclusions

1. Subjects' highest evaluations of Perceived Empathy I occur in Dialogue III (no self-disclosure)
2. Subjects' highest evaluations of Perceived Empathy II occur in Dialogue III (no self-disclosure)
3. Subjects' highest evaluations of Perceived Competence occur in Dialogue III (no self-disclosure)
4. Subjects' highest evaluations of Perceived Trust occur in Dialogue III (no self-disclosure)

Curtis (1981) concludes that there is some doubt about the effectiveness of a therapist's self-disclosure as a therapeutic technique. Additional research is needed to assess the effectiveness of self-disclosure in relation to other theoretical models to see if results would be consistent with those found in this study.

Recommendations

1. Use alternative method of statistical analysis
2. Include better operationalization of constructs; instruments presented with limited reliability statements (and as reference to validity factors)
3. Include procedures for random selection of subjects from population and random assignment of subjects to treatment groups

Quantitative Considerations

Randomly assign clients to therapists who work under the three conditions (high, low, and no self-disclosure). Assess the clients' perceptions of the dependent variables, empathy, trust, and competence.

Qualitative Considerations

The following questions emerge from Curtis (1981) and could drive related qualitative studies:

- What type of therapeutic outcome would the participants project, based upon the vignette read (level of self-disclosure)?
- Why did the participants think they were being assessed? What did they think was the researcher's intent?
- What were the participants' general impressions in response to the therapist in the vignette? What kinds of reactions did they have?
- Based upon previous counseling experiences with a therapist's level of self-disclosure, what were the participants' experiences and expectations of counseling?

Curtis (1981) himself speculates about possible qualitative studies grounded in this research.

Reflections

One may enter the qualitative-quantitative continuum from either a qualitative or a quantitative perspective (refer to figures 1 and 2). One may then critique the study under consideration and suggest future qualitative or quantitative research (symbolically represented by the feedback loops in figure 2). This student critique demonstrates a benefit of our model—it has heuristic value: any particular study leads to other questions and other research.

Critique of the Fuller Study

The third study to which we apply the continuum is "The Monocultural Graduate in the Multicultural Environment: A Challenge for Teacher Educators," by Mary Lou Fuller (1994), found in appendix D.

Neutrality

That the twenty-eight participants were recent graduates from one university (multiple cases, similar time of graduation, and common university experience) would seem to help the researcher diminish bias and build neutrality. Taped interviews and observations were transcribed, lending reduced personal bias to the data as they were analyzed. But the researcher who collected the data also analyzed the data, resulting in a need to monitor personal bias.

Prolonged Engagement On-Site

The goal of prolonged engagement is to take into account distortions in typical experiences of those being studied. The original conceptualization of prolonged engagement grew out of ethnographic studies. The researcher was alerted to such things as being a "stranger in a strange land" (Lincoln & Guba, 1985, p. 302), selective perception, and building trust in the research setting. Fuller's (1994) study, while qualitative, was not an ethnographic study. The researcher engaged in twenty-eight settings rather than one; therefore, the notion of prolonged engagement does not strictly apply.

Conceptually, however, one might conclude that interviewing and engaging in observation of twenty-eight teachers to study the multiple ways individual teachers interact with different cultures is more in keeping with the goal of exploring what is typical (undistorted) in teachers' experiences than including only one or two teachers. In other words, to the extent that twenty-eight provided more evidence for common meaning from research subjects as well as for clarifying and decreasing the potential biases in the researcher's perceptions, this study was strengthened.

Persistent Observation

The object of persistent observation is to achieve depth of meaning from the data (i.e., what seems salient in the setting). Like the other criteria, it was originally described for ethnographic research. To comply with this criterion, the researcher focuses in detail on the most relevant factors in an ethnographic study. The emerging domains of meaning, then, are based on a depth of understanding. To apply this characteristic to the Fuller study (not an ethnographic study) requires examining how the researcher determined what labels to apply to the emerging themes of the teachers' experiences. In this instance, the salience of the themes—that they went beyond superficialities (Lincoln & Guba, 1985)—was substantiated by including within the thematic interpretations only those ideas expressed by at least one-third of the twenty-eight teachers.

Peer Debriefing

No evidence was available.

Triangulation

Because two data-collection methods (observation and interview) were used, triangulation is strengthened. No information is given about the consistency of the data collected between the two techniques. Multiple sources of experience (twenty-eight teachers) are additional evidence of triangulation.

Member Checking

No evidence was available.

Referential Materials

No evidence was available.

Structural Relationships

There is no evidence that the resulting data from the two methods (observation and interview) were compared for structural relationships.

Theoretical Sampling

No evidence was available.

Leaving an Audit Trail

No evidence was available.

Generalizability

The author indicates what is needed for preservice, monocultural teachers who will be teaching in multicultural settings, thus showing her intent to generalize.

Applicability. Although the results do add to the knowledge base about multicultural environments, Fuller limited the study to the cultural environments of schools in Texas, Nevada, and Arizona. The narrative does not provide deep enough description to be able to estimate applicability. In addition, the participants in the study were self-selected. For whatever reason, each chose to teach in his or her location. Those reasons may differ for other teachers to whom others may want to apply the results.

Context limited. There is insufficient information to assess the extent to which the context limits the generalizability. One common feature across the twenty-eight interviews is that the teachers may have been responding in ways that they thought might satisfy the interviewer.

Replicability. There is no way to estimate whether the results would occur again at different times or in different settings.

Negative-Case Analysis

No evidence is provided in the published study that outliers or processes for including them were included in the design.

Truth Value

Truth value, we suggest, is the overall judgment based on all the preceding criteria. The study's limitations in meeting the thirteen criteria suggest a moderate level of truth value.

Reflections

What can be done to add to Fuller's (1994) research from a quantitative perspective? Among many possibilities, a couple of ideas come to mind. The author identifies six categories of meaning about these teachers' multicultural experiences. A survey of items based on these themes could be developed. Ratings from a sample of preservice teachers would allow interrelationships among the themes to be analyzed. Perceptions and experiences could be measured, and the extent to which the six themes are generalizable could be determined. Here, again, the interactive continuum has value in showing how qualitative research can lead to quantitative research, and vice versa. Science, after all, is the ongoing accumulation of knowledge.

Critique of the Rhoades and Kratochwill Study

The fourth study to which we apply another version of the continuum is "Teacher Reactions to Behavioral Consultation: An Analysis of Language and Involvement" by Mary M. Rhoades and Thomas R. Kratochwill (1992), found in appendix E.

Research Question

Do the language of the behavioral consultant and the involvement of the teacher in the behavioral consultation process affect the acceptability of the intervention to be used in regular classrooms?

Independent Variables

- teacher: involvement versus noninvolvement
- language: technical versus ordinary

Dependent Variable

The dependent variable is the acceptability rating for each consultation scenario.

Design in Campbell and Stanley's (1963) Terms

$$\begin{array}{lll} R & X_1 & O_1 \\ R & X_2 & O_1 \\ \hline R & X_3 & O_1 \\ R & X_4 & O_1 \end{array}$$

R = random assignment
X_1 = technical language and teacher involvement
X_2 = technical language and no teacher involvement
X_3 = nontechnical language and teacher involvement
X_4 = nontechnical language and no teacher involvement
O_1 = acceptability rating

Internal Validity

The strengths of this study are that it is a true experimental design and random assignment of teachers to the four groups. Theoretically, then, this design controls for all threats to internal validity: history, maturation, testing, instrumentation, regression, selection, mortality, and interaction of selection and maturation.

External Validity

There are two threats to external validity in the Rhoades and Kratochwill (1992) study: volunteers were used; and there may be reactive arrangements because of the video. The major problem with this design, however, is that it is a simulated design. People respond to videotapes, but we do not know if they would respond the same way with actual encounters. In addition, the subjects were only exposed to twelve minutes of video. In a real-life situation, the exposure would be longer.

Conclusions

In the article, the authors conclude that there are no main effects for involvement and no main effects for language. They should have said that there are no significant main effects. Their stated conclusion implies the acceptance of the null hypothesis, which is scientifically inappropriate.

Qualitative Research Suggestions

As the authors suggest, a qualitative study including interviewing the subjects after they watch the videotapes might identify why some people responded as they did and thereby suggest a quantitative investigation in future studies to make the results more generalizable.

Reflections

It is obvious that this critique is less exhaustive than the critique of Curtis (1981), the other quantitative study. We intended this abbreviated article to show that, even at a broad and general level, use of our continuum model can lend valuable insight into the quality of research.

"Science" and a Search for Principles of Practice in Mixed Methods

The validity in quantitative research is well established; the validity in qualitative research has been subject to long and thorough discussion, and the validity of mixed methods research has only recently begun to provoke serious consideration. For all three paradigms, we argue that consistency among the research purposes, research question, and research methods is the answer to the validity question. Validity (or trustworthiness) is the lens or framework through which "principles" of practice can be constructed, especially for mixed methods research. Principles of practice can increase the likelihood that mixed methods designs are scientifically sound.[1]

This chapter includes

- definition of epistemological consistency
- description of the model of consistency as a set of criteria for mixed methods research
- explanation of a set of beginning principles of scientific mixed methods research

Beginning with concerns for validity led us to this question: Must mixed methods research always be based in a postpositivistic framework? In other words, is it not more legitimate, perhaps, to subsume mixed methods ultimately under quantitative research because of the complicated and contradictory obligations on the mixed methods researcher that require control that may be the antithesis of qualitative researchers' ways of considering the reality being studied?

To serious researchers, mixed methods offered a powerful new paradigm (Tashakkori & Teddlie, 2003). The popularity of mixed methods has blossomed so much lately that some researchers might assume that after constructing domains of meaning from a qualitative study, they can code those themes as "variables," test them empirically, and claim that they are using mixed methods. Unfortunately (or fortunately), it is not that simple because the findings of qualitative studies (e.g., domains of meaning) and the findings of quantitative studies (e.g., probabilistic decisions about hypotheses) have different epistemological assumptions. Mixed methods are extremely valuable but cannot be a panacea (Ridenour & Newman, 2004).

"Principles of Practice" in Mixed Methods Research

Is it possible to develop standards of practice for mixed methods research? Or, are the standards of quantitative research and of qualitative research (reviewed in chapter 3) sufficient to assure the scientific validity of research in education? Morse (2003) has established what she calls "principles" of utilizing mixed methods; and that idea might be a worthy starting point to begin to discuss what might eventually become standards of practice. Rather than "standards," perhaps "principles" is a better place to begin. We agree with Morse.

The *research question* identifies the area of inquiry and the area of interest: What, exactly, is the researcher attempting to find out? Some studies begin with only an inkling of a question, and, some studies begin with a crystal clear, unambiguous question. All researchers begin by questioning whether they are predisposed somehow to carry out qualitative or quantitative studies. Some researchers' thinking begins with only an implicit question; we argue that the question needs to be made explicit. Furthermore, the question needs to be strongly linked to the research purpose.

The *research purpose* is the reason the study is being conducted, the rationale for the study. Multiple purposes can drive a study. The purpose attempts to identify: Why, exactly, is the researcher carrying out this study? For what purposes? First, only through an iterative process of thinking through both the questions and the purposes can each (question and purpose) be optimally clarified (Newman et al., 2003). And, second, only if the questions and purposes are clarified, can the research methods be justified? In other words, a strong justification of the research methods is possible only if the research purpose and research question are explicit, transparent, public, and clearly articulated.

Appendix F presents an expansion on our questions of consistency; it serves as a model of consistency criteria. We present this as a starting point, a set of principles that we believe reflects research as an interactive continuum, each dimension of the study (question, purpose, and method) inextricably linked with the other. For the researcher designing a study, these principles are starting places in making methods decisions, particularly, in this case—about mixed methods:

- The potential research questions must be acknowledged.
- The potential research purposes must be acknowledged.
- Consistency must be justified between the question(s) and the purposes(s).
- Consistency must be justified between the purposes(s) and the method(s).
- Method(s) must be justified as capable of fulfilling the purpose(s).
- Method(s) must be justified as capable of addressing the question(s).
- Epistemological assumptions must be consistent among purpose, question, and methods.

For the consumer of research, these principles are starting points in assessing the scientific quality of the study:

- Methods must be epistemologically consistent with findings.
- Findings must be consistent with methods.
- Purposes must be consistent with conclusions.
- Implications must be consistent with purposes.

This visual model is a weak representation of a process that, like science, is systematic, organized, and sequential. However, the process is dynamic and not always unidirectional, not always linear; rather, it is iterative. It is impossible, unfortunately, to depict a process that we are admittedly convinced is the optimal way to think through research to read scientific decisions about methods. What we are attempting to show is that the researcher thinks through questions and purposes, and after that iterative process, decisions about methods can be legitimately made.

A systematic process of identifying all possible questions and purposes is much more likely to lead to methods that are consistent with *all* the purposes and *all* the questions the researcher is interested in. The scientific quality of a study is strengthened when epistemological consistency exists across all dimensions of the study design. Epistemological consistency is a quality of the systematic procedures of science.

The researcher engages in an iterative process, thinking through all possible research questions and purposes, iteratively aligning them for consistency. Once this process is accomplished, and the questions and purposes are clarified, then methods decisions are made. On this model, we have suggested what might be the "expected" or "traditional" paradigm to address each of the nine examples of purpose (column 3, appendix F). On the one hand, it may well be that one single purpose leads to one single method. On the other hand, a potential, however, is provided here that a richer array of questions and purposes can be acknowledged, and methods aligned with them.

Epistemological consistency means, for example, that a study driven by a purpose to measure change in student learning is carried out through a methodological process that includes valid and reliable measures of learning over time and strategies to calculate change in those measures. Epistemological consistency means also that this same study might have other purposes. And, if there are other purposes (for example, a need to understand how students experienced a change in the instructional environment), epistemological consistency means that the researcher has identified that purpose a priori and incorporated methods to fulfill that purpose, too.

Using the framework of the continuum increases the likelihood that the researcher is using methods that are consistent with the purpose(s) and research questions (Newman & Benz, 1998). The internal consistency among the research question, research purpose, and research method is the basis for beginning to think through these principles for what we prefer to call "holistic" or mixed methods research designs. Perhaps these principles can be a starting point to begin to set standards for the power of mixed methods. Relatedly, in the seventy-fifth anniversary issue of the *Harvard Educational Review*, Egan (2005) recalls a comment by Ludwig Wittgenstein over forty years ago about the state of psychology that seemed to him reflective of contemporary education—a description that calls on contemporary researchers to be much more conceptually clear. The qualitative-quantitative interactive continuum, we think, improves the conceptual clarity and diminishes conceptual fuzziness. To quote Egan (2005), "He [Wittgenstein] characterized the psychology of his day as suffering from a defect that might also be directed at much current educational research: he saw a combination of 'experimental methods and conceptual confusion'" (p. 34).

Standards (or, at least, guiding principles) of practice in using mixed methods research are based on the situation that the contradictions (not the complementariness) of qualitative and quantitative research mean that the challenge to control the research-planning and data-collection strategies are greater than they are in single-paradigm research. Mixed methods research is inherently postpositivist. By definition and by purpose, it has to be. Facing a research situation that requires an integrating of qualitative and quantitative research is the antithesis of the constructivist standpoint of the qualitative researcher. Without a postpositivist epistemology, mixed methods designs would not be feasible. In other words, the researcher needs to a priori identify the purposes, the sources of data or evidence, the means of collecting the evidence, and the plans for analysis. A qualitative, naturalistic, constructivist framework, by definition, must be subordinated to the control needs. The antithesis of a naturalistic attitude is required.

In a mixed methods design, the qualities of science previously discussed require the researcher to consider at least five characteristics or five elements. It is difficult to countenance these five considerations in ways other than an objective (positivist and postpositivist) stance. The researcher must consider

- the assumptions of qualitative research and its validity constraints
- the assumptions of quantitative research and its validity constraints
- the ways in which the qualitative and the quantitative strategies and assumptions inform one another
- the context of the study as far as how the qualitative and quantitative elements serve worthy purposes
- the purpose or purposes of the study itself

To return to the five benchmarks of scientific research, they are processes that are systematic, potentially verifiable, potentially replicable, self-correcting, and carried out for purposes of explanation. We have earlier made the case that quantitative research and qualitative research can be subjected to a set of "standards" that are, at their base, positivistic (Campbell & Stanley, 1963; Lincoln & Guba, 1985). Can mixed methods research designs encompass these five qualities? We have already argued that situations that lead researchers to adopt the mixed methods paradigm are situations that are inherently complicated—more than one set of epistemological assumptions, for example. Therefore, we maintain that the postpositivistic standards apply to mixed methods research designs as they parallel the sequence of evidence collection dictated by the questions and purposes. In other words, as the researcher moves from one set of questions and purposes to another, the standards for evidence collection and analysis shift in a related fashion.

Adopting a basically postpositivistic set of principles for mixed methods research justifies mixed methods research as consistent with the five benchmarks of science. Does this conclusion dilute the naturalistic, constructivist, and idealistic values of qualitative research? We would argue that it does not—science is one avenue toward a knowledge base about education. Science is one way of knowing; it is not the only way of knowing. Other more artistic, naturalistic, liberatory, uncontrolled, and creative ways of knowing about teaching and learning (qualitative research designs) can contribute to a knowledge base.

Some forms of qualitative research (e.g., narrative, emancipatory inquiry, autoethnography) may be more difficult for mixed methods designs to encompass, especially because mixed methods designs are inherently multifaceted and require tighter researcher control than do qualitative studies. This state of affairs does not diminish their value; it merely places them in another category of research. It is not at all problematic that some kinds of qualitative (and, perhaps, quantitative) research may be most appropriately and most advantageously utilized outside a mixed paradigm. Having said this, however, it is impossible to predict what types of quantitative research or qualitative research cannot be "mixed." We are only beginning to question that here; we are speculating here on those kinds of qualitative research that cannot be subsumed within a mixed paradigm because they cannot be aligned with what we have focused on in this current volume—the qualities of science that are inherently in need of control.

In this chapter, we have begun to define terms and raise questions about mixed methods that may lead to a discussion toward principles of practice. Upon a foundation of validity, we have attempted to speculate about science and mixed methods. We have proposed five criteria with which education research must comply to be considered scientific. Standards of validity for both quantitative and qualitative research have opened the possibility of standards of validity for mixed methods research. We have identified three categories of mixed methods research and identified one, the qualitative-quantitative interactive continuum, as the type that we think holds the most promise for being conceptually clear, methodologically consistent, and therefore, possibly more scientific than the other two. Utilizing this type (the interactive continuum), we have shown an iterative process for, first, identifying research questions and purposes and, second, selecting methods of evidence collection and analysis. We have described what has been described by others as the complementary nature of mixed methods and what we see as the complicated nature of mixed methods research and, on those bases, suggested that this third paradigm (mixed methods paradigm) is most effectively viewed through a positivist and/or postpositivist lens. A researcher using mixed methods generally needs to control the design more than do researchers in the other two paradigms. A first consideration of principles of practice have been suggested here—some ideas for us and, we hope, for others to contemplate as a way to strengthen the power of mixed methods in education research by establishing standards of practice.

Appendixes

Notes

Glossary

References

Index

Appendix A: Phenomenological Research—Laughter and Humor

A questionnaire was administered to a small sample of nurses and educators. These samples consisted of females between the ages of twenty-nine and fifty years of mixed marital status and race. Each protocol was separated into meaning units. That is, each response written by each participant was divided into a series of expressions, which, if read consecutively, match the original protocol. Next, each meaning unit was condensed to its central theme. The central themes were combined into the final formal step in this qualitative analysis, namely, the demarcation of the typical components of "why people laugh" and "what laughter does for people" for this sample of nurses and educators.

To summarize, a review of the data allows for a description of the characteristics of the nomothetic characteristics of the phenomena. The nursing and educator groups are united to form a single unit here because the analysis of the protocols renders any differentiation arbitrary at this exploration phase.

In describing why they laugh, the participants in this study indicated a variety of reasons. They seemed to laugh for the sake of laughter and to enjoy life, as well as at something perceived as humorous. Laughter helped lighten stress and covered up less pleasant feelings of sadness and frustration or nervousness and shyness. While some participants experienced laughter as an uncontrollable and spontaneous event, others highlighted their choice to laugh or how they were taught to see the brighter side of a situation. Participants identified a capacity to laugh alone or at self, but some preferred to share laughter, especially the "contagious" type, with others.

The participants in this study emphasized the positive benefits from laughter: they felt good, relaxed, and positive, had unity of mind and emotions, and were glad to be alive. It also provided for an emotional release to ease stress, anxiety, and tension and to cover up nervousness until participants could regain a coping strategy. On the other hand, laughing at inappropriate times could create embarrassment, and extreme laughter could lead to a painful side and nausea (Foerstner, Newman, & Koenig, 1985).

On the following pages are a copy of the questionnaire and examples of typical components of phenomenological experience about why participants laugh and what laughter means (taken from Foerstner et al., 1985).

Results

Typical Components of Phenomenological Experience of Why Participants Laugh and What It (Laughter) Does for Participants in This Study: A Sample

I. Nurses
   A. Why do you laugh?
   - to enjoy life, myself, and friends
   - allows participants to take changes in stride
   - cover up feelings of sadness, frustration, inadequacy, indecision when something is amusing
   - taught to see brighter side of situation, feel happy, and enjoy something
   - people joke, something comical happens, having fun with peers
   - feels good and brings joy into life if blue and down

   B. What does laughter do for you?
   - feel more relaxed
   - eases stressful situation
   - like self better
   - easier to live with
   - feels good
   - light and airy
   - relieves tension and anxiety
   - more positive

II. Educators
   A. Why do you laugh?
   - reflection of feelings, enjoy life
   - at jokes, circumstances, self
   - uncontrolled response to something funny

- spontaneous during conversation from a joke or quick wit of participants or partner
- when a situation makes a giggle rumble inside, and it bursts out

B. What does it do for you?
- feel "up," relaxed, comfortable
- hurts side and feel nauseated if laugh very hard
- releases stress, feels good
- releases energy

Sample of Meaning Units and Central Themes

A: response to "Why do you laugh?"

B: response to "What does laughter do for you?"

I. Nurses

MEANING UNITS	CENTRAL THEMES
A. 1. To enjoy life, myself, and my friends	1. To enjoy life, myself, and friends
2. Laughter/humor lightens stress and can make life (the hard times especially) more easy to adjust to and allows me to take changes in stride.	2. Lightens stress; makes life easier to adjust to; allows participants to take changes in stride
3. Occasionally I laugh because it's easier than crying about a bad situation.	3. Easier than crying about a bad situation
B. 1. Laughter makes me feel more relaxed and eases stressful situation.	1. Feel more relaxed; eases a stressful situation
2. I like myself better when I laugh. I'm easier to live with, and it makes me feel good.	2. Like self better when laugh; easier to live with; makes participant feel good.

II. Educators

A. 1. I laugh because it is a reflection of my feelings. I enjoy life and laugh readily as a result.	1. Reflection of feelings; enjoy life

        2. I laugh at jokes, at circum- 2. Laugh at jokes, circum-
           stances, frequently (very)   stances, self
           at myself.
    B. 1. Laughter makes me feel "up"   1. Feel "up"; feel relaxed; feel
           and relaxed and comfortable.   comfortable
       2. I find it is contagious, and I 2. It is contagious.
           laugh with others, and others
           laugh with me.

To Potential Research Participants:

Your assistance in responding to the attached set of questions on laughter will further our research project into the phenomena of humor. To participate, please fill in the demographic data at the top of the page, and then write out as fully a reply as needed to each question to describe your experience. If, however, for any reason you feel disinclined to participate, please return the questionnaire blank with or without an explanation.

Laughter and Humor

Age _____   Race _____

Sex _____   Education _____

Marital status _____   Occupation _____

Why do you laugh?

What does laughter do for you?

Thank you for your time and assistance.

Appendix B: One Counselor's Intervention in the Aftermath of a Middle School Student's Suicide— A Case Study

Jo Ann C. Alexander and Robert L. Harman

The authors discuss the application of Gestalt theory as a means of dealing with the surviving classmates of a student who committed suicide.[1]

Four young people have died of suicide within the last month in our county. The second of these was Jason, a thirteen-year-old student in the middle school in which I am a counselor. The first death occurred on Valentine's Day, and Jason's followed by three weeks. The subsequent two deaths occurred in other parts of the county within a week of Jason's death. These events seem like a poignant validation of the "Werther" effect (Phillips, 1985)—the tendency of humans to imitate.

It is important for school counselors to have skills not only in programming for suicide but also for intervening in the aftermath of suicide. Existing literature, however, offers little to prepare counselors—particularly those in school settings—for this role. Researchers (Calhoun, Selby, & Faulstich, 1982; Calhoun, Selby, & Selby, 1982) have reported on the aftereffects of suicide, but few actually (Hill, 1984; Zinner, 1987) have discussed the ways in which a counselor might intervene.

In short, I had little from the professional literature to inform me when I learned of Jason's death. My task, as I identified it, was to help our young people grieve over Jason, to assist them in the process of letting go of him, and to minimize the likelihood of copycat suicides. I did not know what to expect in terms of their response to the news. I was coordinator of guidance in the school, in which we had three counselors, one per grade level, and approximately one thousand students. I was assigned to the sixth grade, in which Jason had been a student.

Fortunately, I had been a Gestalt therapy trainee for two years. Also, I had some specific training in working with suicidally depressed adolescents and had done considerable reading in this area. It was with this preparation that I began my interventions. The approach described below should not be used by counselors without the support of comparable theory, knowledge, and skill.

After I decided that the most effective use of my time would be to work primarily with those 150 students with whom Jason had daily contact, I met with the faculty to prepare a consistent and appropriate school-wide response. We agreed not to eulogize Jason but to focus in public on our feelings of grief, shock, loss, fear, and even anger. We would not glorify his act, nor would we ignore that which we would miss about him.

In my work with students, I relied heavily on my knowledge of the theory and practice of Gestalt therapy. My task was to enhance students' awareness of their thoughts, feelings, and sensations about the death of a classmate and also to help them learn to express themselves in ways that might be more nourishing to themselves and to others at this time of trauma. With awareness, students might have more choice about trauma to respond both to Jason's death and to their feelings of isolation, hopelessness, and despair. My intent was to involve each student in his or her present experience in as many ways as possible; I began by visiting six of Jason's seven daily classrooms. Access to these classrooms was not difficult: Six teachers were delighted and relieved to accept my offer to work with their students; only two remained in the classroom to participate in my one class period intervention. One teacher chose to work with his students himself.

Class-sized Groups, Individuals, and Groups of Two

Jason chose to die with his good-byes left unsaid. His act was abrupt and blunt. So as not to deflect from the quality of his act, I entered each classroom and announced, "I'm here today to help you say good-bye to Jason Davis. Jason is dead. He committed suicide. . . . He won't be back. . . . Where did Jason sit?"

Most suicides constitute an unfinished gestalt. In these cases, good-byes are left unsaid, and the question of why a life was taken is left unanswered. Jason's was no exception. The purpose of my work, then, was to encourage students to say good-bye to Jason as a preface to letting go, to experience the collective and individual responses to his death in the here and now, to open avenues for intimate relating, and to explore constructive ways of coping with the situation.

Students acknowledged sadness and anger. My responses were intended to legitimize their feelings of betrayal and resentment. Those who had been the targets for some of his obscure signals were given the opportunity to cry and to speak to Jason's empty seat to tell him of their anger, resentment, betrayal, guilt, grief, confusion, sadness, and emptiness. Also, they were able to tell him what they would have done for him if they had known he was troubled. Others, as well, were given the opportunity to address Jason's empty seat, telling him what they would like him to know. Each was encouraged to end his or her statement with " . . . and good-bye, Jason."

For some, this experience seemed too threatening or overwhelming, so yet another mode of expression was offered: the nonverbal, subvocal good-bye. Students were invited to look at Jason's seat and imagine saying good-bye to him and to imagine telling him what they would like him to know. If time permitted, some classroom groups were given the opportunity to write their good-byes to Jason. The exercise was varied in the art class to allow for another avenue of expression, that of artistic representation of feelings.

Whenever a student exhibited strong emotion, I invited classmates to respond directly to that student. The students were exceptionally kind, caring, and supportive in their relating with each other. Many pleaded with their peers, "Please don't leave me. I'll help you." Others said, "I'm afraid I'll kill myself."

As the day progressed, I noticed that some students had been present in previous classes, so I invited them to remain in the classroom or gave them an opportunity to go to the library instead. Only one child, Jason's closest school friend and classmate in all of his seven classes, elected to go to the library. He did, however, choose to participate in five classroom sessions and requested two additional sessions, one of which is described in the section below.

At the conclusion of these classroom sessions, I offered the opportunity for additional counseling. As a result, several students sought individual or dyadic (students in pairs) sessions. Because of an expressed continuing need, I formed a small group that met once and another small group that met weekly for the remaining three months of the school year. The individual and dyadic sessions, as well as the small-group sessions, were similar to the work done in the classrooms but with more intense, focused attention.

Small-Group Sessions

Initial Group

The group sessions proved to be by far the most intense of the counseling sessions that I conducted. I employed with these students a projective technique adapted from that suggested by Oaklander (1978). Students were asked to think for one minute about Jason and his death. When time was called, they each were provided with a huge sheet of paper and some crayons, and they were asked to express their feelings on paper in colors, lines, shapes, and symbols. I paid attention to how each student approached and continued the task as well as to the picture itself. This proved valuable in helping students to reown previously disowned parts of themselves and to identify some who currently might be considered at risk.

The drawing of one student, Jason's closest school friend, seemed very simple and resembled the letters "JOI." In speaking as though he were each part of his drawing, he described his own feelings of emptiness, loneliness, and confusion, as well as his own suicidal fears:

> This is the part of my brain that says, "Do it."
> This is the part of me that says, "Don't do it."
> I'm a hook with a sharp end. I can hurt you.
> I'm going round and round.
> I'm left hanging. I'm empty inside.
> I'm straight and bright and happy when I don't think about Jason.

A portion of our subsequent time was spent on his belief that he must keep himself busy so that he would not think about Jason. His fear was that if he thought about Jason, he might hurt himself. Consequently this child was expending a tremendous amount of energy in his attempts not to acknowledge feelings and was experiencing a great deal of anxiety. In the group setting, he was able to express his feelings in a safe environment and to receive caring and support from group members.

Subsequent Group

As a result of several students' expressed continuing need, I formed a long-term group and held weekly sessions for the remaining three months of the school year. This group was composed of six girls, three of whom had been in the initial group and wanted to continue. Because some closure had been reached for the other two members of the initial group, I formed another group, which was to meet for twelve weeks. During the first session, Lynne was observed tearing pieces from her notebook as she spoke about her sadness and confusion. In the Gestalt mode of staying with "what is," Lynne was invited to continue to tear her notebook and to see where that might bring her. When asked to give her hand a voice, she said, "I'm tearing up my notebook in little pieces." When asked if there was anyone in her life she would like to tear up, she replied, "Yes. Jason and me."

I directed her to "tear Jason up, tell him how you feel about his leaving you." After she completed her response to Jason, I invited her to become the pieces and give them a voice, at which time she described being "all torn up, broken, nothing but a pile of pieces.... I should have done something to stop him. I knew. It's my fault. I hate myself."

Lynne seemed to be making progress on undoing her process of retroflecting (turning back onto herself) her anger and destruction when Anna tearfully interrupted, "This is the second time this has happened to me. My brother committed suicide." At this point, the focus of the group's attention turned to Anna.

During Anna's intensive work on her brother's suicide, many of her comments suggested that she believed her peers were laughing at her, thinking she was dumb. So that she could become aware of what was really out there, I invited her to look at each person and tell me what she observed. She reported seeing each person looking at her and not laughing, but she was still imagining she was dumb.

I told her to "look at each person." At this point, she perceived much genuine warmth from the group members. In addition to the support being given to Anna, each student was now voluntarily holding hands with one or two other group members. The group ended with each girl looking directly at one or more of the other group members and clearly stating what she needed from that other person. Many said, "Don't be my friend and leave me like Jason did."

Many of our subsequent sessions proved to be as intense. During our third session, four of the six members revealed that they had attempted suicide. Two reported at least two prior attempts. The remaining two reported having seriously considered suicide.

Conclusion

I do not, unfortunately, know the actual impact of my work with these students. I do know, however, that I was deeply moved by their capacity for grieving and for caring for each other. Through our work together, I developed a great deal of caring for these young people, whose behavior had previously not drawn me to them.

Several themes emerged from my encounters with Jason's classmates. They were experiencing the various grief responses and a pronounced fear that others would follow Jason's example. Not only were they greatly afraid being faced with the loss of yet another friend but many also feared their own suicidal potential. The incidence of previous suicide attempts was alarming. The death of a friend highlighted several other issues as well: poor self-concept, excessive self-demands, fear of loss, grief over previous losses, self-blame, and self-recrimination.

Although this is not a study of the responses of teachers, administrators, and counselors in the school, my observation is that they feel unprepared to deal effectively with such a tragedy. Perhaps, consequently they are prone to avoid the issue. In this case, they seemed shocked and almost paralyzed. Most wanted someone else to handle the situation.

Generally, the students who characteristically exhibited such problem behaviors as skipping school and disrupting class were the most verbal participants. These students seemed "stirred up" by Jason's suicide. They were the risk takers again but this time in a positive and healing way. They were the catalysts who brought their classmates together in more intimate, supportive, and caring ways.

Note

1. Reprinted from Alexander, J. A. C., and Harmon, R. L., One counselor's intervention in the aftermath of a middle school student's suicide: A case study, *Journal of Counseling and Development*, 1988, 66, 283–85. © 1988 The American Counseling Association. Reprinted with permission. No further reproduction authorized without written permission from the American Counseling Association.

References

Calhoun, L., Selby, J., & Faulstich, M. (1982). The aftermath of childhood suicide: Influences on the perception of the parent. *Journal of Community Psychology, 10,* 250–54.

Calhoun, L., Selby, J., & Selby, L. (1982). The psychological aftermath of suicide: An analysis of current evidence. *Clinical Psychology Review, 2,* 409–20.

Hill, W. (1984). Intervention and postvention in schools. In H. Seidak, A. Ford, & N. Rushforth (Eds.), *Suicide in Young* (pp. 407–15). Littleton, MA: John Wright.

Oaklander, V. (1978). *Windows to our children.* Moab, UT: Real People.

Phillips, D. (1985). The Werther effect: Suicide and other forms of violence are contagious. *Sciences, 25,* 32–39.

Zinner, E. (1987). Responding to suicide in schools: A case study in loss intervention and group survivorship. *Journal of Counseling and Development, 65,* 499–501.

Appendix C: Effect of Therapist's Self-Disclosure on Patients' Impressions of Empathy, Competence, and Trust in an Analogue of a Psychotherapeutic Interaction

John M. Curtis
Los Angeles, California

Summary—The present study examined the relationship between a therapist's self-disclosure and the patients' impressions of the therapist's empathy, competence, and trust. Written dialogues were constructed to manipulate three conditions of high, low, and no disclosure by the therapist. Fifty-seven subjects were randomly selected and assigned to one of three treatment conditions, and the Barrett-Lennard Relationship Inventory and Sorenson Relationship Questionnaire were measures of perceived empathy, competence, and trust. Findings confirmed the initial predictions: the greater the use of therapist's self-disclosure, the lower the subjects' impressions and evaluations of the therapist's empathy, competence, and trust. The results raise doubt regarding the predictability of therapist's self-disclosure as a psychotherapeutic technique and suggest that, at least with respect to the type of self-disclosure used in this study, therapists who utilize self-disclosing techniques may risk adversely affecting essential impressions on which a therapeutic alliance is established.

The use of therapist's self-disclosure has been part of the current controversy regarding the distinctions between counseling and psychotherapy. This controversy presumably began with Rogers (1951) who, in an attempt to deemphasize the "medical model" influence borrowed from psychoanalysis, coined the term counseling to characterize more appropriately the psychotherapeutic endeavor.

Classical psychotherapeutic technique, which presumably originated with Freud (1912/1958), contraindicated the utilization of therapist's self-disclosure; instead, therapist's anonymity, that is, the "blank screen" or "mirror" posture, and personal restraint (the "rule of abstinence") were recommended to help mitigate the contamination of the patients' transference reactions.

This preliminary caution was corroborated by many other psychoanalytic theorists and clinicians (Fenichel, 1941; Glover, 1955; Greenson 1967; Langs, 1973, Menninger, 1958) by indicating that the therapist's expressed personal reactions tended to interfere with the analysis of the patients' transference discoveries and resolutions.

The emergence of non-psychiatric specialties as providers of psychotherapeutic service, a growing discontent with the genetic of psychoanalysis, as well as its limitations in terms of time, expense, and narrow range of patients to whom the treatment was applicable, and the escalating influence of behavioristic and humanistic-existential psychology, led to the development of new psychotherapeutic techniques.

In a marked departure from the traditional "blank screen," psychoanalytic posture, several theorists and researchers have identified what they deem to be essential therapeutic determinants: (a) Rogers (1957), the attitude of congruence; (b) Jourard (1964), who coined the term self-disclosure, the attitude of transparency; (c) Bugental (1965), the attitude of authenticity; (d) Kaiser (1965), the attitude of openness; and (e) Truax and Carkhuff (1967), the attitude of genuineness.

Several investigations (Davis & Skinner, 1974; Gary & Hammond, 1970; Jourard & Resnick, 1970; Worthy, Gary, and Kahn, 1969) substantiate what Jourard (1971) designates as a "dyadic effect" of self-disclosure: that self-disclosures offered by the first party in a dyadic interaction elicit self-disclosures in the second party.

Other studies (Dies, 1973; Feigenbaum, 1977; Jourard & Friedman, 1970; Murphy & Strong 1972; Vondracek & Vondracek, 1971) have shown that the use of therapist's self-disclosure favorably influences clients' perceptions necessary to the development of a strong therapeutic alliance; these findings, of course, are consistent with a humanistic-existential and/or behavioristic perspective.

In contradistinction to the aforementioned studies, however, are yet other investigations (Chaikin & Derlega, 1974a, 1974b; Derlega et al., 1976; Polansky, 1967; Truax et al., 1965; Vondracek, 1969; Weigal et al., 1972; Weigal & Warnath, 1968) which show that therapist's self-disclosure has adversely affected clients' impressions, that is, perceptions and evaluations, regarding the therapist's "mental health" and professional comportment. These findings, of course, are consistent with the established recommendations of a psychodynamic orientation (cf. Freud, 1912/1958; Fenichel, 1941; Glover, 1955; Greenson, 1967;

Langs, 1973; Menninger, 1958) and display noticeable equivocality in the available research.

Since the use of therapist's self-disclosure represents a salient distinction between psychodynamic and humanistic-existential paradigms of psychotherapy and because the inconsistencies found in previous research raise doubt as to the predictability of its [therapist's self-disclosure] technique, further research has become necessary.

A major objective of the present study was to help test the effectiveness of therapist's self-disclosure as a psychotherapeutic technique by achieving increased control over the independent variable. To accomplish this, following the research precedents of Rogers (1957) and Truax and Carkhuff (1967) on facilitative therapeutic conditions, the dependent variables of (a) empathy, (b) competence, and (c) trust were selected for investigation. These variables have been recognized to be essential inferences on which a therapeutic alliance is established, irrespective of the clinician's theoretical orientation.

It was hypothesized ($p < .05$) that (1) the therapist's *high* self-disclosure condition will yield the patients' lowest evaluations of empathy, competence, and trust; (2) the therapist's *low* self-disclosure condition will yield the patients' moderate evaluations of empathy, competence, and trust; and (3) the therapist's *no* self-disclosure condition will yield the patients' highest evaluations of empathy, competence, and trust.

Method

Subjects

Fifty-seven subjects currently receiving psychotherapeutic services, twenty-nine of whom are male and twenty-eight of whom are female, whose combined mean age was thirty-two years and ranged between eighteen and fifty-five years, were randomly selected from a metropolitan mental health and family treatment agency. Subjects were asked to participate in a study designed to enhance services for patients to which they were randomly assigned to one of three conditions; these included three levels of therapist's self-disclosure (high, low, and none), the only experimental manipulation.

Design

A one-way analysis of variance design (see Kerlinger, 1973), employing three randomized groups of subjects, which corresponds to the conditions of therapist's high, low, and no self-disclosure, was utilized.

Measures

Independent variable. Three written dialogues between a patient and therapist were constructed to serve as the stimulus conditions in which therapist's self-disclosure was manipulated across three separate conditions, that is, the dialogues were specifically designed to vary systematically the corresponding levels of therapist's high, low, and no self-disclosure.

To accomplish this with optimal control over the independent variable, the patients' comments in all three dialogues were held constant, while only the therapist's remarks were varied systematically to reflect the change in the level of self-disclosure. Moreover, in order to assure this control the content of the therapist's remarks was held constant across all three dialogues; and only the form was altered by carefully changing the pronoun used in the therapist's responses.

Specifically, the pronoun *I* was utilized in the therapist's responses in Dialogue I (e.g., I sometimes feel depressed too), the therapist's high self-disclosure condition. The pronoun *we* was utilized in the therapist's responses in Dialogue II (e.g., we all sometimes feel depressed), the therapist's low self-disclosure condition. The pronoun *it* was utilized in Dialogue III (e.g., it must have made you feel depressed), the therapist's no self-disclosure condition.

Further, to camouflage the subtle differences displayed in the three dialogues and to establish greater uniformity between the dialogues, three of the eight responses by the therapist were held constant by inserting three uniform, nonrevealing, that is, reflective, responses. Specifically, in Dialogue I, five of the therapist's eight responses included direct personal references; in Dialogue II, five of the therapist's eight responses indirect personal references; and in Dialogue III, all eight of the therapist's responses included only reflective, non-revealing responses.

Dependent variables. The Sorenson Relationship Questionnaire is a twenty-four-item counseling-relationship questionnaire, using seven-point Likert-type scales for measuring the therapist's performance in the counseling situation (A. G. Sorenson, *Toward an instructional model for counseling*. Occasional Report No. 6, Center for the Study of Instructional Program, University of California, Los Angeles, CA, 1967). The instrument is divided into two parts, that is, part I measures empathy and Part II measures expertness. For the purposes of this study, both scales were utilized. Internal consistency reported was .85. The Barrett-Lennard Relationship Inventory (1962) is a sixty-four-item counseling-relationship inventory, using six-point Likert-type scales for measuring the four counselor/therapist traits of (a) regard, (b) empathy, (c) unconditionality, and (d) congruence. For the purposes of this study, only the empathy and unconditionality scales were utilized. Unconditionality, as defined by Barrett-Lennard (1962), was taken to represent a measure of trust. Split-half reliability reported as .82 (Barrett-Lennard, 1962).

Procedures

Subjects were randomly selected from a file of ongoing treatment cases and then randomly assigned to one of the three experimental conditions, that is, the written dialogues. The subject's therapist was notified regarding the researcher's intention to enlist his patient in the study; written consent was obtained, and appointment times were scheduled.

The experimenter greeted each subject individually in the agency's waiting room just prior to his scheduled appointment time. The subjects were then asked if they would participate in a study being conducted by the agency to enhance services to patients. Upon agreeing to participate, the subjects were presented individually with the pre-selected dialogue and the two relationship questionnaires. Next, subjects were instructed to read the dialogue and obtain a general impression of the therapist's performance, and to evaluate this performance by completing the two distributed questionnaires. Subjects were also informed that, one week following the completion of their questionnaires, they would be invited to discuss any questions related to the research. This served to motivate the subjects toward conscientious participation. Subjects were debriefed as necessary.

Data Analysis

The data analysis included means and standard deviations of the dependent variables. Also, one-way analyses of variance (Myers, 1966) employing the Newman-Keuls multiple comparison procedure (Winer, 1962)—a *post hoc* test of mean differences—was utilized to analyze the sets of data.

Results

The means and standard deviations of the subjects' perceptions and evaluations of the therapist's relative levels of (a) two measures of empathy, (b) competence, and (c) trust are shown in table 1.

Table 1. Means and Standard Deviations of Perceived Therapist's Empathy, Competence, and Trust under Three Levels of Therapist's Self-Disclosure

Measure of Perceived Therapist		Self-Disclosure		
		High	Low	N
Empathy*	M	29.68	30.05	37.73
	SD	6.23	7.27	9.37
Empathy II**	M	56.94	67.10	72.93
	SD	14.18	12.27	11.71
Competence†	M	58.57	67.42	75.52
	SD	13.20	10.10	8.67
Trust‡	M	30.57	31.36	38.63
	SD	7.76	8.59	8.45

* Barrett-Lennard's Empathy Scale of the Relationship Inventory (Barrett-Lennard, 1962).
** Sorenson's Empathy Scale (Relationship Questionnaire, part 1).[1]
† Sorenson's Expertness Scale (Relationship Questionnaire, part 2).[1]
‡ Barrett-Lennard's Unconditionality Scale (Relationship Inventory, Barrett-Lennard, 1962).

The one-way analysis of variance performed on Empathy I (see table 2) yielded an F of 6.21 ($p < .01$). A multiple-comparison procedure of mean differences utilizing the Newman-Keuls post hoc test indicated that this effect was based on the differences between the following conditions: (a) a therapist's high self-disclosure and none ($p < .01$) and (b) a therapist's low self-disclosure and none ($p < .05$). The subject's highest evaluations of Empathy I, as predicted, occurred with self-disclosure by a therapist.

Table 2. Analyses of Variance of Therapist's Perceived Empathy I[†] and II[‡]

Source	df	Empathy I		Empathy II	
		MS	F	MS	F
Between	2	392.75	6.21*	1216.64	15.53*
Within	54	63.23		78.34	
Total	56				

[†] Barrett-Lennard's Empathy Scale of the Relationship Inventory (Barret-Lennard, 1962).
[‡] Sorenson's Empathy Scale (Relationship Questionnaire, part 1).
* $p < .01$.

The one-way analysis of variance performed on Therapist's Perceived Empathy II (see table 2) yielded an F of 15.33 ($p < .01$). A multiple-comparison procedure of mean differences utilizing the Newman-Keuls post hoc test indicated that this effect was based on the differences between the following conditions: (a) therapist's high self-disclosure and no self-disclosure ($p < .01$) and (b) a therapist's low self-disclosure and no self-disclosure ($p < .01$). The highest evaluations for Empathy II, as predicted, occurred with no therapist's self-disclosure.

The one-way analysis of variance performed on Therapist's Perceived Competence (see table 3) yielded an F of 7.38 ($p < .01$). A multiple-comparison procedure of mean differences utilizing the Newman-Keuls post hoc test indicated that this effect was based on the differences between therapist's high self-disclosure and none ($p < .01$). The subjects' highest evaluations for Competence, as predicted, occurred with no self-disclosure by a therapist.

Table 3. Analyses of Variance of Therapist's Perceived Competence[†] and Trust[‡]

Source	df	Competence		Trust	
		MS	F	MS	F
Between	2	1365.11	7.388	374.36	5.17*
Within	54	184.92		72.32	
Total	56				

[†] Sorenson's Expertness Scale Relationship Questionnaire, part 2.
[‡] Barrett-Lennard's Unconditionality Scale (Relationship Inventory, Barrett-Lennard, 1962).
* $p < .01$.

The one-way analysis of variance performed on Therapist's Perceived Trust (see table 3) yielded an F of 5.17 ($p < .01$). A multiple comparison procedure of mean differences utilizing the Newman-Keuls post hoc test indicated that this effect was based on the differences between the following conditions: (a) high therapist's self-disclosure and no therapist's self-disclosure ($p < .05$) and (b) low therapist's self-disclosure and no therapist's self-disclosure ($p < .05$). The highest evaluations for Trust, as predicted, occurred with no therapist's self-disclosure.

Discussion

One of the specific findings of this study was that therapist's self-disclosure adversely affected the subjects' impressions of the therapist's empathy, competence, and trust: the *higher* the level of therapist's self-disclosure the *lower* the subjects' evaluations of the therapist's performance on these prescribed dimensions.

The results do not confirm data from a number of investigations (Davis & Skinner, 1974; Dies, 1973; Jourard & Jaffe, 1970; Nilsson, Strassberg, & Bannon, 1979; Vondracek & Vondracek, 1971) which have indicated that therapist's self-disclosure may have significant therapeutic value, both in terms of its beneficial effects on clients' impressions and on ultimate therapeutic outcome. Rather, the results run somewhat parallel to other studies (Chaikin & Derlega, 1974a, 1974b; Derlega et al., 1976; Polansky, 1967; Truax et al., 1965; Vondracek, 1969; Weigal et al., 1972; Weigal & Warnath, 1968) which have demonstrated unfavorable effects of disclosure by the therapist on clients' perceptions and attitudes.

Since clients' impressions influence the "dyadic effect" of self-disclosure, that is, a reciprocal transaction process between counselor and client (see Jourard, 1971), it would be reasonable to expect that the clients' negative impressions of the therapist would unfavorably affect this exchange process. Further, in a number of studies which seem to corroborate a "dyadic effect" (Davis & Skinner, 1974; Gary & Hammond, 1970; Jourard & Resnick, 1970; Worthy, Gary, & Kahn, 1969) it is difficult if not impossible, to determine whether the effect of self-disclosure per se, or some other aspect of the interpersonal communication, such as similarity of attitude leading to liking, is partly responsible for some of the effects obtained. For instance, similarity of attitude (see Rubin, 1973) leads to liking which, along with the effect of a positive first impression (see Jones et al., 1968), affects subjects' inferences, that is, attitudes and evaluations, on a variety of measures, such as, likeability, professionality, rapport, "mental health," etc. Moreover, even Jourard and Jaffe (1970) recognized that research on self-disclosure is particularly sensitive to these "halo effects" which require increased control and precision. This was accomplished in this study by systematically varying the level of the therapist's self-disclosure in written dialogues.

Other recognized sources of potential invalidity to which research on self-disclosure is especially vulnerable are: (a) experimenter bias (see Rosenthal, 1966), a factor to which Jourard and Jaffe (1970) attribute many of the reported effects; (b) selection factors (see Campbell & Stanley 1963) for how different samples, such as, undergraduate college students or impatient schizophrenics, affect predictions; and (c) demand characteristics (i.e., the role of subjects' expectations) such as a subject's preconceived attitudes which influence the outcome. These potentially contaminating factors were controlled in this study by utilizing written dialogues as the treatment conditions, and by selecting subjects currently receiving psychotherapeutic services—the group to which generalizability was targeted—as participants.

The present study was analogue in nature and cannot be regarded as an actual therapeutic encounter. The results might have proven somewhat different had subjects, having long-term affective involvement characteristics of a psychotherapeutic relationship, been evaluating their own therapist's performance. This potential limitation, however, needs to be assessed against the benefits obtained from increased experimental control.

Since the therapist's self-revealing responses are uniformly contraindicated by psychodynamically oriented treatments and since the results seem to be consistent with these established recommendations, these findings need to be interpreted in light of the specific type of therapist's self-disclosure used in the study, that is, the change in the pronoun used in the therapist's remarks, which corresponds to high, low, and no disclosure by the therapist. Culbert (1970) identified several types of therapist's self-disclosure, such as, anecdotal, experiential, feeling responses, etc., which, when incorporated into similar experimental conditions, may have yielded somewhat different results.

The fact that reciprocity has been recognized as a powerful determinant of dyadic interaction (see Homans, 1961; Jourard, 1971) and that a therapist's self-disclosure has been used to promote this kind of transaction cannot be regarded as conclusive evidence supporting its therapeutic utility. Aside from its appropriateness in "everyday" nontherapeutic, social interaction, the use of therapist's self-disclosure has been found in this and other research (Chaikin & Derlega 1974a, 1974b; Polansky, 1967; Vondracek, 1969) to yield unpredictable results.

Since the effect of first impressions (Jones et al., 1968) influences subjects' perceptions and evaluations and since the therapist's self-disclosure may promote patients' undesirable inferences of empathy, competence, and trust, this behavior may interfere with the establishment of rapport in a therapeutic relationship. At the very least, the effects of therapist's self-disclosure—if predictability may be regarded as a measure of utility—cannot yet be predictably determined across its many varieties and conditions.

Clearly, additional research is necessary, perhaps incorporating alternative varieties of therapist's self-disclosure—anecdotal, experiential, feeling responses, etc.—and contrasting effects between men and women, among different diagnostic groups, as well as between short- and long-term therapy. Such research may be helpful in determining the effects of therapist's self-disclosure under these prescribed conditions.

Note

Reproduced with permission of author and publisher from: Curtis, J. M., Effect of therapist's self-disclosure on patients' impressions of empathy, competence, and trust in an analogue of a psychotherapeutic interaction, *Psychological Reports*, 1981, 48, 127–36. © Psychological Reports, 1981. This research was supported in part by a grant from the Beverly-Linden Mental Health Foundation.

References

Barrett-Lennard, G. T. Dimensions of therapist responses as causal factors in therapeutic change. *Psychological Monograph*, 1962, 76, No. 4 (Whole No. 562).

Bugental, J. T. *The search for authenticity*. New York: Holt, Rinehart & Winston, 1965.

Campbell, J. T., & Stanley, J. C. *Experimental and quasi-experimental designs for research*. Chicago: Rand McNally, 1963.

Chaikin, A. L., & Derlega, V. J. Looking for the norm-breaker in self-disclosure. *Journal of Personality and Social Psychology*, 1974, 42, 117–29. (a).

Chaikin, A. L., & Derlega, V. J. Variables affecting the appropriateness of self-disclosure. *Journal of Personality and Social Psychology*, 1974, 42, 588–93. (b).

Culbert, S. A. The interpersonal process of self-disclosure: It takes two to see one. In R. T. Golembiewski & A. Blumberg (Eds.), *Sensitivity training and the laboratory approach*. Itasca, NY: Peacock, 1970, pp. 88–100.

Davis, J. D., & Skinner, A. E. G. Reciprocity of self-disclosure in interviews: Modeling of social exchange. *Journal of Personality and Social Psychology*, 1974, 29, 779–84.

Derlega, V.J., Lovell, R., & Chaikin, A. L. Effects of therapist self-disclosure and its perceived appropriateness. *Journal of Consulting and Clinical Psychology*, 1976, 44, 886.

Dies, R. R. Group therapist self-disclosure: An evaluation by clients. *Journal of Counseling Psychology*, 1973, 20, 344–48.

Feigenbaum, W. M. Reciprocity in self-disclosure within the psychological interview. *Psychological Reports*, 1977, 40, 15–26.

Fenichel, O. *The psychoanalytic theory of neurosis*. New York: Norton, 1941.

Freud, S. Recommendations to physicians practicing psychoanalysis. In *Standard edition of the complete psychological works of Sigmund Freud*. London: Hogarth Press, 1958. Vol. 12. pp. 109–20. (Originally published, 1912).

Gary, A. L., & Hammond, R. Self-disclosure of alcoholics and drug addicts. *Psychotherapy: Theory, Research, and Practice*, 1970, 7, 151–71.

Glover, E. *The technique of psychoanalysis*. New York: International Universities Press, 1955.

Greenson, R. R. *The technique and practice of psychoanalysis*. New York: International Universities Press, 1967.

Homans, G. C. *Social behavior: Its elementary forms*. New York: Harcourt, Brace & World, 1961.

Jones, E. E., Rock, L., Shaver, G. K., & Ward, L. M. Pattern of performance and ability attribution: An unexpected primacy effect. *Journal of Personality and Social Psychology*, 1968, 10, 317–40.

Jourard, S. M. *The transparent self*. Princeton, NJ: Van Nostrand, 1964.

Jourard, S. M. *The transparent self*. (Rev.). Princeton, NJ: Van Nostrand, 1971.

Jourard, S. M., & Friedman, R. Experimenter-subject "distance" and self-disclosure. *Journal of Personality and Social Psychology*, 1970, 15, 278–82.

Jourard, S. M., & Jaffe, P. E. Influence of an interviewer's self-disclosure on the self-disclosure behavior of interviewees. *Journal of Counseling Psychology*, 1970, *17*, 252–57.

Jourard, S. M., & Resnick, J. L. Some effects of self-disclosure among college women. *Journal of Humanistic Psychology*, 1970, *10*, 84–93.

Kaiser, H. *Effective psychotherapy*. New York: Free Press, 1965.

Kerlinger, F. S. *Foundations of behavioral science research*. (2nd ed.). New York: Holt, Rinehart & Winston, 1973.

Langs, R. *The technique of psychoanalytic psychotherapy*. Vol. 1. New York: Jason Aronson, 1973.

Menninger, K. A. *Theory of psychoanalytic technique*. New York: Basic Books, 1958.

Mowrer, O. H. *The new group therapy*. Princeton, NJ: Van Nostrand, 1964.

Murphy, K. C., & Strong, S. R. Some effects of similarity self-disclosure. *Journal of Counseling Psychology*, 1972, *19*, 121–24.

Myers, J. L. *Fundamentals of experimental design*. Boston: Allyn & Bacon, 1966.

Nilsson, D. E., Strassberg, D. S., & Bannon, J. Perceptions of counselor self-disclosure: An analogue study. *Journal of Counseling Psychology*, 1979, *26*, 399–404.

Polansky, N. A. On duplicity in the interview. *American Journal of Orthopsychiatry*, 1967, *37*, 568–80.

Rogers, C. R. *Client-centered therapy*. Boston: Houghton Mifflin, 1951.

Rogers, C. R. The necessary and sufficient conditions of therapeutic personality change. *Journal of Consulting Psychology*, 1957, *21*, 95–103.

Rosenthal, R. *Experimenter effects in behavioral research*. New York: Appleton-Century-Crofts, 1966.

Rubin, Z. *Liking and loving. An invitation to social psychology*. New York: Holt, Rinehart & Winston, 1973.

Truax, C. B., & Carkhuff, R. R. *Toward effective counseling and psychotherapy: Training and practice*. Chicago: Aldine, 1967.

Truax, C. B., Carkhuff, R. R., & Kodman, F. Relationships between therapist-offered conditions and patient change in group psychotherapy. *Journal of Clinical Psychology*, 1965, *21*, 327–29.

Vondracek, F. W. The study of self-disclosure in experimental interviews. *Journal of Psychology*, 1969, *72*, 55–59.

Vondracek, S. I., & Vondracek, F. W. Self-disclosure in pre-adolescents. *Merrill-Palmer Quarterly*, 1971, *17*, 51–58.

Weigal, R. C., Dinges, N., Dyer, R., & Straumfjord, A. A. Perceived self-disclosure, mental health, and who is liked in group treatment. *Journal of Counseling Psychology*, 1972, *19*, 47–52.

Weigal, R. C., & Warnath, C. F. The effect of group psychotherapy on reported self-disclosure. *International Journal of Group Psychotherapy,* 1968, *18,* 31–41.

Winer, B. J. *Statistical principles in experimental design.* New York: McGraw-Hill, 1962.

Worthy, M., Gary, A. L., & Kahn, G. M. Self-disclosure as an exchange process. *Journal of Personality and Social Psychology,* 1969, *12,* 59–63.

Appendix D: The Monocultural Graduate in the Multicultural Environment—A Challenge for Teacher Educators

Mary Lou Fuller
University of North Dakota

I always thought I would teach in the Midwest. I had my whole life planned and those plans certainly didn't include teaching in Texas (Kim). Kim imagined herself teaching in an elementary classroom similar to those of her childhood. She would teach white, middle class children in a Midwest town or a suburb. But Kim and others graduating from colleges of education will find the classrooms in which they teach strikingly different from those of their childhoods. In this article I report on a study exploring what happens when monocultural students live and teach in multicultural environments and draw implications from those experiences for teacher education faculty and curricula.

Kim will find the class she teaches much more diverse from any she experienced as a student. By the year 2000 between 33% (Commission on Minority Participation in Education, 1988) and 40% (Hodgkinson, 1989) of school children will be from ethnically/racially/culturally/economically diverse groups. Increased immigration, higher birth rates for minorities, and a declining number of white non-Hispanic children are the causes of these demographic changes (Griffith, Frase, & Ralph, 1989). Currently, the majority of students are children of color in twenty-three of the nation's twenty-five largest cities (Gay, 1989; National Center for Educational Statistics, 1987).

Kim will also find the economic status of her students' families may vary considerably from that of her family. Haberman (1989) predicts that one third will live in poverty by the year 2000. Kim will find the impact of poverty exacerbated because this population includes a large proportion of children six and younger—the most developmentally sensitive years—(Children's Defense Fund, 1989), with even higher proportions for Hispanic and Black children (Moneni, 1985).

Teacher education faculty must recognize the new demographics and identify and respond to their educational implications. They cannot assess the effectiveness of their professional practices without considering the needs of contemporary classrooms and teachers. An important criterion of the effectiveness of teacher education faculty is the teaching proficiencies of their graduates, an effectiveness seen by examining the ways education graduates relate to the changing school environment and population.

Little evidence of change in teacher preparation or teachers' classroom strategies exists, despite marked demographic changes in classrooms, a situation Sleeter and Grant describe as *business as usual* (1993, p. 18). They review how teachers teach diverse populations, what they teach, and how they group students; they find educational changes incongruent with demographic changes. Schools generally do not meet the needs of children from diverse populations.

The population of the public schools is changing, but that of colleges of education is not. Teacher education programs generally serve students coming from largely middle class homes (Webb & Sherman, 1989) with middle class sensibilities. The numbers of white, female, and middle class preservice teachers are increasing (Fuller, 1992a; Zimpher & Ashburn, 1992; and Webb & Sherman, 1989), while the number of teachers from diverse populations is decreasing (Zapata, 1988). Thus the teaching population is becoming more monocultural (Hodgkinson, 1989; National Education Association, 1987), while the student population is becoming more multicultural.

These primarily white, female, middle class preservice teachers from rural or suburban environments (Fuller, 1992a) have little exposure to people different from themselves. They also experienced little contact with minorities in the colleges of education they attended as these institutions generally reflect the same demographics as the students' communities. Though one might expect that the preservice curriculum would attempt to remediate the students' multicultural deficiencies (Mills & Buckley, 1992), with few exceptions, this is not the case (Zeichner, 1993; Fuller 1992a, 1992b; Gwaltney, 1990; Griffith, Frase, & Ralph, 1989; Kniker, 1989).

Preparing preservice teachers for their future classrooms grows more complex as the school population becomes more diverse. Changing demographics require changing teacher education strategies; education faculty must consider the demographics of their graduates' classrooms and inform themselves of their graduates' experiences in these new environments. Documenting what happens as monocultural preservice teachers (students with limited exposure to diverse populations) begin teaching in multicultural settings was a goal of this study.

I explored the experiences of monocultural elementary education graduates teaching in multicultural environments; I looked for the accommodations of graduates of a Midwestern elementary education teacher education program working and living in environments different from those they experienced as children. The university is in a primarily monocultural area, and its students are similar to preservice teachers nationally (Zeichner, 1993; Fuller, 1992a; Kniker, 1989).

In the study I view one of the major goals of multicultural education as: . . . *reform [of] the school and other educational institutions so that students from diverse racial, ethnic, and social class groups will experience educational equity* (Banks, 1994, p. 3). Based upon this definition, I distinguish those who are *multicultural*—well informed about and have had meaningful experiences with people of diversity—from those who are *monocultural*—those who have not had meaningful experiences with diverse populations. I affirm the philosophical position of social reconstructionism which . . . *prepares future citizens [e.g., as preservice teachers] to reconstruct society so that it better serves the interest of all groups of people. This approach is visionary. Although grounded very much in the everyday world of experience, it is not trapped in this world* (Sleeter & Grant, 1993, p. 210).

Methods and Data Sources

Teachers in the study were recent elementary education graduates from an upper Midwestern university. All had taken one of two courses: Multicultural Education or Introduction to Indian Studies. The Multicultural Education course is anthropological in nature, provides information about given cultures, develops an understanding of the concepts necessary to work with children from other cultures, and is social reconstructionist in philosophy. Approximately ¾ of the teachers interviewed in this study had taken Multicultural Education. They also studied multicultural concepts in the Introduction to Education course which all had taken.

I contacted all graduates within three years of the study's start and screened them to determine if they were teaching in culturally diverse communities. I used the following criteria to determine three geographic areas to visit in this study: the presence of diverse cultural groups; the presence of small towns, suburbs, and cities; and a number of graduates teaching in the area.

I used two data collection methods, the first a broad protocol including many open-ended questions to interview the teachers. I taped all these interviews as well as those with personnel directors and administrators in the districts employing the teachers. In the latter interviews I sought their impressions of monocultural teachers from the upper Midwest. The second data collection procedure was field observations of the teachers in action and the school community. I tape recorded and transcribed these observations. The classroom observations and interviews of school personnel were to verify the accuracy of the teachers' self-reports. I identified recurring themes and considered important those which a third or more of the teachers mentioned (Slotnick, 1982). I noted less frequently cited ideas and experiences providing particular insights and used participants' voices to clarify issues and provide examples.

Participants

I contacted and screened 354 elementary education graduates to determine their geographic distribution and professional responsibilities. Ninety-one percent (321) responded, all but 29 being involved in professional activities. Respondents lived in twenty-nine states and four foreign countries.

I selected teachers in three states to interview and observe. I visited four communities in central Texas, a metropolitan area in Nevada, and a metropolitan area and two rural areas in Arizona. I wrote letters to all teachers and followed up with phone calls requesting permission to interview them and observe their classrooms. I arranged to interview administrators and personnel directors in the same districts.

I interviewed twenty-eight teachers and observed twenty-six teaching in their classrooms (schedule conflicts prevented the observation of the other two). All of the participants grew up in monocultural communities and identified their families as middle class. All grew up in intact families; one reported her parents divorcing during her late teens. Three were male; twenty-five, female. Five taught kindergarten; six, first grade; three, second grade; one, third grade; one, fourth; one, fourth-fifth combination; two, fifth grade; two, seventh; one, eighth; four, special education; and one was a substitute teacher. I also interviewed five personnel directors and two elementary school principals.

Themes

Analysis of the transcripts reveals six themes: reasons for relocating to their present teaching environment; satisfactions related to the relocation; dissatisfaction related to the relocation; recognition of personal growth; advice to preservice teachers; and feelings about their move.

Reasons for the Move

Forty-two percent of the participants originally preferred to stay in the upper Midwest but moved elsewhere *because that's where the jobs were*. Twenty-nine percent moved to the Southwest because of personal relationships and another 29% because they wanted new experiences. Although they moved to a part of the country more culturally diverse than their home environment, none reported cultural diversity as a consideration in moving to the Southwest.

The following comments reflect the experiences and feelings the participants shared concerning their moves to the Southwest. Jennie (all names are pseudonyms) had not even considered leaving the Midwest until she acknowledged the reality of the job market. *I didn't pick Texas, Texas picked me. I thought for sure I was going to get a job in Minneapolis—after all I had applied. Late in the summer I went to Chicago to apply and on their computer they had three screens full of people with my last name* (Olson) *and there were three other applicants with my exact name, Jennie Olson. The moment of truth! School had already started and I had resigned myself to subbing for the year when I got a job offer in Texas.*

For Nancy, the move to the Southwest was even more serendipitous. *I didn't intend to be here. I went to the Career Fair, standing in what I thought was a long line for positions in the Twin Cities* [Minneapolis and St. Paul] *and when I got to the table I found I was in the line for a large* city district in the Southwest. *After standing in line all that time I thought I might as well interview. As the summer wore on I became desperate—I hadn't heard from anyone. Then one day* [the large city district] *called my home and talked to my mother who volunteered that while she was sure I was professionally organized my bedroom at home was always a mess. Needless to say I was surprised when they called back and offered me a teaching position. After arriving it only took me a few weeks to know I was where I should be.*

Those who moved to the Southwest because of personal relationships were female; relationships varied from married to engaged and other configurations. Sandy's husband's position dictated where she would teach: *My husband graduated in Airport Administration and applied in a variety of cities. It was agreed that I would apply wherever he got a position. I'm surprised, but I really like it here.* Mary Jane, by contrast, moved to pursue a relationship. *I did my student teaching here because this is where my boyfriend was living and I figured that this was the best way to get a teaching position near him. I'm glad that I did because I really like it here although the job turned out to be much more interesting than the relationship.*

Several moved to the Southwest for new experiences. Neither culture nor geographic location was important in deciding where to live. Karen wanted to be away from home and independent: *I wanted new experiences; I wanted to go anywhere. When I grew up in* [a small town], *I just wanted to get out of Dodge big time. I just wanted to know what the rest of the world was like and teaching not only gave me a way to leave, it gave me permission to leave. It was something my parents could understand. After all the money they spent on my education, they wanted to see me employed.*

Satisfactions

Moving was just the beginning; once there, participants began accommodating to both teaching and the new environment. The second theme concerns teachers' satisfaction as a result of moving to the Southwest. The participants reported more satisfactions, both in number and quality, than any other theme. Professional satisfactions included their individual schools (86%), the cultural diversity of schools and communities (75%), their school districts (64%), and the school personnel (64%). Satisfactions in their personal lives included cultural diversity

(75%), more things to do (64%), weather (54%), nice people (50%), living conditions (46%), shopping (42%), beauty of the locale (39%).

Professional Satisfactions. The people the participants met after their moves provided important satisfaction. One participant said, *They just think we [teachers from Midwest] are so great and we just think they are so friendly, warm and nice.* Interviews with administrators and personnel directors confirmed that they saw the participants as great teachers.

Participants were generally enthusiastic about people they met: *I love the people, I like my peers, my principal—she is wonderful, and the parents have been wonderful too* (Candy). *People down here are very nice. If you need questions answered they are always willing to help* (Glenn).

Cultural diversity was important. Margaret commented: *I think that I've grown up a lot because of my move. I've gained a lot more experience than I would have if I had stayed in [state] just basically because I'm teaching a group of kids that are a lot different than I am. I am constantly learning things from them. I'm finding out all kinds of interesting things about different cultures. . . . Their different holidays, and their lifestyles, what is important to them and their families. It is interesting to see some of the things that the parents I'm talking to see as important compared to what my parents viewed as important.* Jayne observed: *My classroom is about half black and half white. Also I have a little girl who is from the Philippines and lives with her mother who is deaf and a little boy from Korea who has limited English proficiency. It is a wonderful class and I think the diversity is part of what makes it so great.*

Sandy noted that appreciating another culture gave her a greater appreciation of her own, and she learned to see herself in different ways: *When you consider the characteristics of someone else's culture you can't help but think about your own.*

Personal Satisfactions. Participants reported satisfaction with the activities available in their personal lives, *the more moderate temperatures, the kindness of people, the living conditions, shopping, and the beauty of the locale.* One of the underlying factors in this category was the number of options available: many things to do, different places to go, many things to buy, and a range of people to meet.

Their personal lives were enriched in a variety of ways. Kathy's experience was representative of many. Initially anxious about her move, she later expressed a high degree of satisfaction with her choice. *I was so nervous, I had never been so far from my family but I fell in love with it as soon as I got here. I walked in the front door of the school and Bobbi [principal] gave me a hug and I felt that I was home. The staff became my family and the teacher in the next room my best friend.* Living conditions, shopping, beauty of locale were also important: *Going to buy clothes and having more than one option. Going out for the evening and having lots of choices of things to do. Meeting more and different kinds of people. Always having more than one opportunity whether it be for fun, necessities, or jobs. I guess the key word is choice* (Mary Jane).

Dissatisfactions

The third theme dealt with dissatisfactions which appeared in fewer areas and which participants reported less frequently than satisfactions. They identified five common concerns: poverty (52%), gangs (50%), lonesomeness for their families (46%), child abuse (42%), and their students' family problems (39%). They mentioned three other dissatisfactions specific to given areas: participants' personal safety (42%); the theme that *you can't trust people the way you can back home* (36%) [reported almost exclusively in urban areas, typically by those recently moving to the area]; extensive and inappropriate standardized testing [only teachers in Texas] (36%).

Professional Dissatisfactions. The participants believed cultural diversity enriched their lives, but other forms of diversity—particularly in their professional lives—disturbed them. Prominent among these were economic diversity (poverty) and diverse family structures (e.g., single parent homes). While not mentioning Maslow's hierarchy, participants repeatedly discussed poverty and family structures in terms of their students' hierarchy of needs. They talked of unmet physical needs, such as hunger, safety needs focusing primarily on the unpredictability of some students' lives, and affiliation needs often resolved through gang membership.

The participants were unprepared for the poverty they saw. All but eight had children of poverty in their classrooms in numbers ranging from a few to most children. Many teachers, like Carol, initially had difficulty recognizing poverty. *I was surprised when a number of my second grade students came to school hungry. They came to school with a lot of problems, but not enough to eat. I just wasn't prepared for that and at first I didn't even recognize what was going on.* Amber had difficulty recognizing the educational implications of poverty. *Sometimes I felt annoyed when my great lessons were met with a lack of enthusiasm and then I realized that it wasn't a lack of enthusiasm but rather a lack of energy. I had to remind myself that it is hard for kids to concentrate on math when their basic needs haven't been met.*

While sensitive to their students' needs, participants lacked understanding of poverty and spoke impatiently about the parents who were victims of poverty. Their observations often suggested *blaming the victim* (Ryans, 1971); some expressed the belief that if *they* just tried harder, *they* could overcome economic adversities. They generally lacked understanding of the nature and causes of poverty, but they did not lack concern for their students in poverty. They often expressed that concern as a desire to help by being there. Amber observed: *I think that you make more of an impact on the lives of these children than on the lives of more affluent children.*

The teachers' childhood familial structures differed dramatically from those of their students. They recognized this and reported that their lack of knowledge of and experience with different family structures constrained their understanding of the students' families. *An area of diversity, at least for me, is the various family structures in my classroom. Very few of the children in my room live with their original parents. They live with single moms, one with a single father, stepparents, a couple live with their grandparents, and one divides her time between her mother and father's home. They don't look like families that I grew up with or know. I have to keep reminding myself that just because they don't look like my family doesn't make them any less of a family. But these families do seem to have a lot more difficulties* (Kathy). The teachers had particular difficulty separating complications in students' lives caused by limited income rather than by family style. They often saw single-parent homes as the cause of a child's problems when in fact they were describing difficulties created by a lack of resources.

Gangs were a problem to varying degrees in all but one district and represented another professional concern. The participants were uniformly unschooled on this topic. Nancy was still trying to understand when she observed: *It affected me right away. During the new teacher meetings I was told that I couldn't wear a bandanna to school because gangs wear bandannas. My students can't wear anything with LA Raiders emblems on them because they are also gang symbols . . . and my students threaten one another with their big brothers who have gang affiliations. My* students are only in second grade and we aren't an inner city school. Those who had inservice training on this topic were much more sophisticated in their understanding of gangs. Vicky and a few others considered issues such as why gangs are attractive to children. She observed: *I think that the attraction of the gangs is the element of belonging. I don't think they* [gang members] *belong to a family or have any real friends. I don't think that they see themselves as fitting in anywhere so consequently gangs are very attractive to them.*

Personal Dissatisfactions. Each personal item reflects an adjustment to a new environment: lonesome for their families, concerned about their personal safety, and feeling that they *Can't trust people the way you can back home.* Participants used their home environment as reference points in making sense of their new environment. While she talked of trust, Jill's tone was one of incredulity as she repeated the same story three times: *There are a lot of nice people here but then there are some that are not so nice. [Another issue is . . .] being trusted. You have to have a bank card to get a check cashed! Everything is checked and then checked again. Can you believe that I had to be fingerprinted twice to teach in this district?* Check-cashing cards are uncommon in her state.

Personal Growth

Another important theme was personal growth. With obvious pride and pleasure, participants talked often about growing and changing as a result of their moves; all but one saw the Southwest's cultural diversity as important to their growth. Sixty-eight percent reported greater independence than they would have had if they had not left home; 64% believed they were now much more open to new experiences. While cultural diversity was not an issue in moving to the Southwest, the participants most often cited it as contributing to personal growth. While one might expect those participants with the most children of color in their classrooms to be most affected by cultural diversity, all but two discussed cultural diversity at some length. *I feel more open and accepting of people. You lose your prejudice and you learn to be accepting of one another's differences* (Glenn). *I have really been dependent upon my parents even though I didn't live at home. I lived on campus but they were only a hop, skip, and a jump away. I've missed them a lot more than I thought I would. I've learned to be very independent and to make my own decisions* (Amanda). *I was ready to come. I like living somewhere larger so I think I've adjusted well. It was a growing experience. I think that I've grown more in the last eight months than I have at any other point in my life. I started a brand new job, I finished college, I moved into a new apartment, left my family and friends, and did it all in one week. I think that I adjusted well. I love it, I really do. There are times, though, that I miss my family* (Toni). Their growth was evident in their resentment of stereotyping of their students and their families. As they became better informed, they became increasingly resentful of cultural misconceptions. For example, many of the participants reported that folks back home would ask questions indicating they had preconceived, negative attitudes toward people of other cultures. Interestingly, the comments they reported resenting are not uncommon among monocultural preservice teachers. However, now that the participants were living in culturally diverse settings, they uniformly resented these comments and questions. Toni noted: *Someone said to me, 'I suppose they steal and all that stuff.' My Euro-American parents view me as a babysitter but my Mexican families see me as someone special; they call me teacher as a form of respect. They are really a gentle culture and not like what you see on TV where they are portrayed as thieves.*

Twenty-six of the twenty-eight participants attributed growth primarily to cultural diversity. Their interviews and classroom behaviors support their assessments. One participant appeared genuinely unaffected by the cultural diversity in her environment; another did not understand its significance.

Jackie, who was unaffected, had created in her very multicultural community an environment closely duplicating her upper Midwest roots. After many job interviews, she accepted a position in an upper middle class, white school and moved into an exclusively white, upper middle class apartment complex. She did not travel outside her part of the community or experience the cultural diversity her city offered. Jackie also expressed discomfort with the culturally diverse schools, people, and neighborhoods in her city. Her classroom mirrored perfectly the upper Midwestern school where she student taught except for the two non-English speaking students who were bused from another part of the city and were invisible in her classroom. She vaguely smiled in their general direction, but she did not acknowledge them in any other way, did not make eye contact with them, did not call them by name or speak directly to them during the several hours she was observed teaching. Nor, predictably, did the other students. Jackie did not provide these students—or their classmates—with the opportunity to . . . *draw people into a public place* (Brown, 1992, p. 8); she did not allow inclusion (Brown, 1992) for the two Hispanic students in her class. Jackie established a monocultural niche in her multicultural community and so missed the growth other participants experienced. Jackie showed that for change to occur, both the presence of cultural diversity and the desire to experience that diversity are required. Just being there is not enough.

Terri also missed the significance of life in a multicultural environment. Only two of the students in Terri's classroom spoke English as their first language and many others were first generation Americans. Her classroom was pleasant, but nothing in it or in Terri's teaching strategies reflected the cultural background of her students. Asked about the cultural characteristics of the students, Terri insisted there were no differences between her present students and those in her (upper Midwest) student teaching experience. She did not see the cultural differences. She likes her students; unfortunately, she is not culturally sensitive.

Advice for Preservice Teachers

The last two themes emerged in response to specific questions. First, I asked the participants what advice they had for a preservice teacher who might want to teach in a classroom next to theirs. Though they offered a variety of suggestions, they mentioned only six frequently enough to be noted: Learn Spanish (93%); take field experiences in multicultural environments (82%); take the Multicultural Education course (78%); take the Classroom Management course (57%); learn about families (50%); and be open minded and flexible (46%). Concerning learning Spanish, even those participants who had no Spanish speaking students believed it important. John said: *I would tell everyone to take Spanish. Spanish has become the second language in this country and it will help you personally and certainly help you professionally.* Jayne noted the importance of multicultural education on finding herself in a town in Texas: *I remember thinking to myself, OK, now what did I learn in Multicultural Education? What am I supposed to do? The first thing I remembered was this is their culture and I must respect it. I need to respect them and their culture and not try to use my culture as a yardstick for other people's behaviors. Then I remembered [the professor] saying over and over that it was my responsibility to be well informed about my students and their culture. I still had some big adjustments to make but I felt secure that I could do it.*

Of the six items, five had preservice curricular implications while the other item concerned personal growth. Interestingly, while everyone discussed concerns about their students' families, only half saw this as an area in which the students could prepare themselves. Similarly, the majority of the interviewees expressed concern about poverty and gangs and yet no one mentioned either subject for study. Evidently, the participants did not see these topics as subjects for inclusion in a teacher preparation curriculum.

Feelings about the Decision to Move

All twenty-eight said they were pleased with their decisions to move to the Southwest. Twenty-one (75%) agreed without any qualifications; six (21%) said that they might move home at some point in time; one person observed that it took her about eighteen months before she felt pleased with her life in the Southwest.

Several of the teachers have moved since arriving in the Southwest, and several others were planning to move. All of the moves had been or were anticipated to be within the Southwest. In general, those contemplating moves were single and were moving to enhance their social lives.

The Six Themes in Summary

From the six themes (reasons for moving, satisfactions, dissatisfactions, personal growth, advice to preservice teachers, and assessment of their moves to the Southwest) emerged a pattern indicating the participants viewed positively their moves to more diverse areas of the country. Although ethnic/racial/cultural diversity was not a factor in their decisions to move to a multicultural community, it became the single most important element of their experience. They generally functioned very well in their classrooms and grew personally and professionally. Their professional concerns included: lack of knowledge and experience with diverse populations, diverse family structures, poverty, gangs, and child abuse. Since these are all curricular issues, thoughtful teacher education faculty members can address them.

The participants' personal satisfactions included pride in their independence and growth, enjoyment of the friendliness of new colleagues and acquaintances, and the more relaxed work environment. The personal dissatisfactions (missing family, concern for safety among those living in metropolitan areas, and feelings of mistrust) are not topics faculty can effectively address.

Educational Implications

The participants believed that the multicultural environments enhanced their personal and professional lives. While they were appreciative of and sensitive to the cultural diversity in their classrooms, my observations suggest that their teaching strategies were generally not culturally informed. They had profited from their preservice multicultural courses but lacked the ability to select appropriate teaching strategies for the environments. Preservice programs must provide the necessary information and should include multicultural field experiences. Such field work will provide the frame of reference preservice teachers need. It is difficult to internalize specific strategies when one lacks experiential reference. The participants themselves identified the need for more preservice multicultural field experiences.

College of education programs must expand their views of diversity to include the effects of social class and families. First, the curriculum should include issues such as poverty, family study, and additional multicultural training (including field experiences as noted) to better prepare students to work in multicultural environments. This will provide preservice teachers with the broad view notion of diversity (Zimpher & Ashburn, 1992) necessary to prevent parochialism.

Second, preparing preservice teachers for their move from a monocultural and small city environment to a multicultural and metropolitan one will ease their transitions. These two changes, taken together, will allow new teachers to focus their professional concerns sooner on the children they teach.

Failure to implement these changes will be very costly to the educational system generally and to the education of children of color particularly. Elementary teachers increasingly represent mainstream society, and they are generally not well prepared to teach diverse student populations. To counter this problem, colleges of education must simultaneously continue to recruit preservice teachers of diversity while carefully preparing the primarily white, middle class teachers who are and will continue to be the teachers for most children of color.

The education of white, middle class teachers must prepare them to understand and appreciate diversity, as well as identify and use the appropriate instructional strategies for their students. Only in these ways will schools be able to address the needs of the diverse populations they serve. Failure to do this, in Villegas' view (1991), will result in serious difficulties: *It seems clear from the research that unless teachers learn to integrate the cultural patterns of minority communities into their teaching, the failure of schools to educate children will continue* (p. 19).

Consequently, preservice teachers must be knowledgeable in the strategies that best meet the academic needs of their students and for colleges to better prepare preservice teachers for diversity. The following is a list of specific strategies effective teachers of minority students use (Irvine, 1992):

- Have appropriately high expectations for students;
- Employ many different instructional materials and strategies;
- Use interactive rather than didactic methods;
- Use the students' everyday experiences in an effort to link new concepts to prior knowledge;
- Help students become critical thinkers and problem solvers.

Zeichner (1993) suggests the following bearing on preparation of preservice teachers:

- Students are helped to develop a clearer sense of their own ethnic and cultural identities;
- Students are taught about the dynamics of prejudice and racism and about how to deal with them in the classroom;
- Students are taught about the dynamics of privileges and economic oppression and about school practices that contribute to the reproduction of societal inequalities;
- Students are taught various procedures by which they can gain information about the communities represented in their classrooms;
- Students complete community field experiences with adults and/or children of another culture;
- Students live and teach in a minority community (immersion);
- Instruction is embedded in a group setting that provides both intellectual challenge and social support.

Implementing these changes to the preservice curriculum should make graduates such as those interviewed in this study more attuned to the needs of and approaches to all their students. The result will be teachers able to both help their students and themselves. They may build on the insight Jennie expressed so well: *It would be so boring if all of my students were white and middle class. I'm so glad that I'm here. Moving made me look past myself at other people and other cultures.*

Note

Reprinted with permission. Copyright by the American Association of Colleges for Teacher Education. Fuller, M. L., The monocultural graduate in the multicultural environment: A challenge for teacher educators, *Journal of Teacher Education*, 45(4), Sept.–Oct. 1994: 269–77.

References

Banks, J. A. (1994). *Multi-ethnic education*. (3rd ed.). Boston: Allyn Bacon.

Brown, C. E. (1992). Restructuring for a new America. In M. E. Dilworth (Ed.), *Diversity in teacher education: New expectations*. San Francisco: Jossey-Bass.

Children's Defense Fund. (1989). *A vision for America's future*. Washington, DC: Children's Defense Fund.

Commission on Minority Participation in Education and American Life (1988). *One third of a nation*. Denver: American Council on Education and Education Commission of the United States.

Fuller, M. L. (1992a). Monocultural teachers of multicultural students: A demographic clash. *Teacher Education*, 4(2).

Fuller, M. L. (1992b). Teacher education programs and increasing minority school population: An educational mismatch. In C. A. Grant (Ed.), *Research directions for multicultural education: From margin to mainstream*. London: Falmer Press.

Gay, G. (1989). Ethnic minorities and educationally equality. In J. A. Banks & C. A. Banks (Eds.), *Multicultural education: Issues and perspectives* (pp. 167–88). Boston: Allyn Bacon.

Griffith, J. E., Frase, M. J., & Ralph, J. H. (1989). American education: The challenge of change. *Population Bulletin*, 44(4), 16.

Gwaltney, C. (1990). Almanac: Facts about higher education in the nation, the states, and D.C. *The Chronicle of Higher Education*, 11–29.

Haberman, M. (1989). More minority teachers. *Phi Delta Kappan*, 71(10), 771–76.

Hodgkinson, H. L. (1989). *The same client: The demographics of education and services delivery systems.* Washington, DC: The Institute of Education.

Irvine, J. J. (1992). Making teacher education culturally responsive. In M. E. Dilworth (Ed.), *Diversity in teacher education: New expectations.* San Francisco: Jossey-Bass.

Kniker, C. R. (1989). *Preliminary results of a survey of Holmes and non-Holmes group teacher education programs.* Chicago: Midwest Holmes Group.

Mills, J. R., & Buckley, C. W. (1992). Accommodating the minority teaching candidate: Non- black students in predominantly black colleges. In M. D. Dilworth (Ed.), *Diversity in teacher education: New expectations* (pp. 134–59). San Francisco: Jossey-Bass.

Moneni, J. A. (1985). *Demography of racial and ethnic minorities in the United States: An annotated bibliography with a review essay.* Westport, CT: Greenwood Press.

National Center for Educational Statistics. (1987). Washington, DC: U.S. Printing Office.

National Education Association. (1987). *Status of the American public school teachers, 1985–1986.* Washington, DC: National Education Association.

Ryans, W. (1971). *Blaming the victim.* New York: Vintage Books.

Sleeter, C. E., & Grant, C. A. (1993). *Making choices for multicultural education.* (3rd ed.). New York: Merrill.

Slotnick, H. B. (1982). A simple method for collecting, analyzing, and interpreting evaluative data. *Evaluation in the health professions,* 5(3), 245–58.

Villegas, A. M. (1991). Culturally responsive pedagogy for the 1990s and beyond. *Trends and issues paper No. 6.* Washington, DC: ERIC Clearinghouse on Teaching and Teacher Education.

Webb, R. B., & Sherman, R. R. (1989). *Schooling and society.* (2nd ed.). New York: Macmillan Publishing Company.

Zapata, J. (1988). Early identification and recruitment of Hispanic teacher candidates. *Journal of Teacher Education,* 39, 19–23.

Zeichner, K. M. (1993). *Educating teachers for cultural diversity.* East Lansing, MI: National Center for Research on Teacher Learning.

Zimpher, N. L., & Ashburn, E. A. (1992). Countering parochialism in teacher candidates. In M. D. Dilworth (Ed.), *Diversity in teacher education: New expectations* (pp. 40–59). San Francisco: Jossey-Bass.

Appendix E: Teacher Reactions to Behavioral Consultation—An Analysis of Language and Involvement

Mary M. Rhoades and Thomas R. Kratochwill
University of Wisconsin–Madison

Explored [are] two dimensions of behavioral consultation that can potentially influence teachers' reactions to the consultation process. Two independent variables (consultee involvement and consultant language) were completely crossed to create four videotape scenarios differing only with respect to the manipulated variables. Elementary school teachers ($N = 60$) were randomly assigned to view and rate one of the four scenarios on a measure of acceptability. Subjects reported high ratings for technical language when the psychologist took a directive role and did not involve the teacher in the problem-solving process. Results are discussed within the context of previous acceptability research and future research concerning consultee involvement in the consultation process.

To provide services effectively to the greatest number of students, school psychologists often focus their intervention efforts on the teacher through behavioral consultation (Bergan & Kratochwill, 1990). Research has indicated that behavioral techniques are often effective, but that they are sometimes misinterpreted and evaluated by teachers as less acceptable than other techniques (Elliott, 1988). The link between acceptability and effectiveness has been demonstrated in recent research, and practitioners are recognizing the need to explore consumer reactions when implementing interventions (Reimers, Wacker, & Koeppl, 1987). Consumer acceptability of psychological interventions is also important from a legal and ethical standpoint in establishing the social validation of a technique and in understanding individual needs.

Despite attempts to make behavioral strategies more acceptable to consultees and more efficient in managing classroom difficulties, consulting school psychologists still encounter teacher resistance as a major obstacle to effective service delivery (Witt & Elliott, 1985; Witt & Martens, 1983). Two psychologist-mediated variables may have a potential effect on teacher acceptability. These variables include the degree to which the teacher is *involved* in a collaborative problem-solving process with the consultant and the amount of *technical language* used by the consultant in communication with the teacher (Elliott, 1988).

Researchers investigating why the bias against behavioral techniques exists, or how it might be changed, have not presented conclusive results, but have suggested a number of variables that should be considered. Woolfolk, Woolfolk, and Wilson (1977) found that college students viewing identical videotapes of teaching strategies rated those strategies labeled "behavior modification" as less effective than videotapes labeled "humanistic." A second study examined whether the presentation of a rationale for efficacy or a softening of behavioral terms would influence ratings (Woolfolk & Woolfolk, 1979). Results indicated that video-taped behavioral teaching strategies presented with rationales and humanized terms were rated favorably, but only by an undergraduate group. The behavioral language of the technique did, to some degree, determine raters' preferences, but the actual bias against behavioral teaching strategies was not clearly defined.

Medway and Forman (1980) expanded the work of Woolfolk et al. (1977) to the area of consultation by presenting videotapes of mental health and behavioral consultation between a school psychologist and a teacher to actual teachers and school psychologists in the field. Results of this study indicated that although psychologists preferred the mental health technique, teachers rated the behavioral model as more effective.

A series of three experiments using written case descriptions of teaching methods evaluated by undergraduate students was completed by Kazdin and Cole (1981). The potency of the labeling effect was questioned as a causal variable in the negative evaluation of behavior modification techniques. It was shown that the negative evaluation received by the behavior modification condition, as compared to humanistic and neutral conditions, was due primarily to the content of the method, rather than the label applied. It was suggested that past research into the negative evaluations of behavioral techniques may have misplaced emphasis on the importance of the label alone and that other variables be examined.

The impact of labeling bias was explored by Witt, Moe, Gutkin, and Andrews (1984). Case descriptions of classroom interventions evaluated by teachers indicated that a pragmatic description was rated as more acceptable than behavioral or humanistic alternatives. The behavioral description, emphasizing that "staying in at recess involved the contingent application of punishment for the explicit purpose of controlling the child's inappropriate behavior" (Witt et. al., 1984, p. 364), was evaluated as least acceptable, especially when rated by more experienced teachers. It appears that the acceptability of equivalent interventions may be at least partially determined by the language used in the approach.

A review of research in this area indicates a potential bias against behavioral techniques, but only two studies used actual subjects from the field (Medway & Forman, 1980; Witt et al., 1984), and most focused on teaching techniques as opposed to school-based consultation (Kazdin & Cole, 1981; Woolfolk & Woolfolk, 1979; Woolfolk et al., 1977). Studies completed by Kazdin and Cole (1981) and by Witt et al. (1984) suggest a need to explore separately the issues of content and technical language in evaluating the bias against behavioral techniques. A stronger recommendation for such analyses could be made if future work in the field with videotaped scenarios, as opposed to written case descriptions, supported such hypotheses.

Active involvement of the consultee in the consultation process may be perceived as important because the teacher is in a unique position to provide perspectives on the utility of the intervention and possibly, ownership of intervention plans by the teacher will facilitate intervention integrity (Gutkin & Curtis, 1990; Witt, 1990). Reinking, Livesay, and Kohl (1978) reported that consultee implementation of programs developed during consultation are related directly to consultee involvement. It is often concluded that consultees prefer collaborative, rather than expert, consultation styles (Babcock & Pryzwansky, 1983; Fine, Grantham, & Wright, 1979; Wenger, 1979) and that solutions developed through collaborative consultation are more acceptable than those generated alone or by others (Fairchild, 1976; Reinking et al., 1978). But results are not clear cut. For example, Wenger (1979) examined teacher responses to a consultant's attempt to facilitate either a collaborative or an expert consultation relationship. The collaborative consultant involved the teacher in the process of determining the child's needs and in developing strategies and techniques for classroom interventions. Although the expert consultant condition included teacher involvement (e.g., input, perceptions, hypotheses), the consultant developed the recommendation and gave them to the teacher. The teachers exposed to the collaborative approach were more satisfied (as measured by ratings), but there were no significant findings on the recommendation implementations. The design was also quasi-experimental.

In the case of the Babcock and Pryzwansky (1983) study, three groups of education professionals (elementary school principals, special education teachers, and second-grade teachers) rated their preference for four consultation models as offered by school psychologists at five stages of consultation. The professionals rated the collaboration model the highest on a rating scale. But, finding a stage by model interaction led these authors to conclude that consultation preference should not be considered a unidimensional concept.

More recently Wiese and Conoley (1989) reported that perceptions of problem-solving style and exposure to a collaborative or expert consultation model did not significantly affect consultee self-reported problem-solving behaviors or expectations regarding a problem solution. Ratings of acceptability were used following undergraduate viewing of one of two videotapes depicting the two consultation conditions.

Involvement has also been examined from the perspective of a relational communication analysis. Erchul (1987) assessed consultants and consultees on two measures of relational control, domineeringness, and dominance. His results indicated that consultants controlled the dyadic relationship across all stages of behavioral consultation. Moreover, consultants having high dominance scores tended to be judged as more effective by consultees. Results of the study by Erchul and Chewning (1990) also indicated that in dyads where the consultant is dominant and the consultee is submissive, consultation outcomes are considered more positive by both parties. These authors suggest that behavioral consultation consists of a more cooperative than collaborative relationship.

To provide further information on the complexity of the consultant-consultee relationship, we were interested in examining two variables that have been studied in various contexts and dimensions in previous research, but not in behavioral consultation. Specifically, we assessed consultee involvement and consultant language in a completely randomized design for their effects on a measure of acceptability.

Method

Subjects

Participants included sixty white regular education teachers from fifteen public elementary schools (K-6) in a Midwestern metropolitan area of approximately 170,000 people and three smaller elementary schools in three rural communities.[1] Each school enrolled three hundred to five hundred students, employed sixteen to twenty-five full-time teachers, and included at least one special education program for students with exceptional learning needs. Only teachers with access to school psychological services at their schools were included. Pupil services staff working on-site at each school included the school psychologist, social worker, and guidance counselor. Subjects included fifty-three females and seven males and half had at least sixteen years of teaching experience. None of the teachers had training in behavioral consultation. Most participants were either relatively new or long-time veterans at their present school. Teachers who had been teaching in their present school placement for one to three years made up 27% of the subject sample while 37% of the subjects had been teaching at their present school for sixteen years or more. Participants were adequately distributed across each grade placement as follows: $K = 3$, $1 = 13$, $2 = 9$, $3 = 9$, $4 = 15$, $5 = 6$, and $6 = 5$.

Procedures

After district approval to conduct research in the schools had been obtained, elementary principals were contacted by letter and phone to request school participation. When participation was secured, teachers were contacted by letter and asked to volunteer approximately thirty minutes of their time to participate in the viewing and rating of a videotaped consultation intervention. Consenting teachers were then scheduled for a viewing at their convenience.

Subjects were assigned randomly to one of four conditions: technical language with teacher involvement, technical language without teacher involvement, nontechnical language with teacher involvement, or nontechnical language without teacher involvement. Participants viewed one of the four taped scenarios of behavioral consultation between a female school psychologist and a female classroom teacher.

Prior to viewing the tape, subjects were given a brief introduction to consultation and their role in assessing the consultation scenarios. Teachers were told that the videotape they were about to view depicted a process for identifying and analyzing a problem and devising a plan for dealing with the child's problem in the classroom. They were also informed that other teachers would view a tape with a different style of consultation with the psychologist. Teachers were told that they should rate the acceptability of the consultation procedure and not the specific behavioral plan for the child portrayed in the tape. Copies of this introduction were given to each subject and were also read aloud by the experimenter. After viewing the consultation scenario, teachers were asked to complete the IRP-15 (Martens, Witt, Elliott, & Darveaux, 1985) which included a Likert rating of fifteen items designed to assess acceptability of the consultation procedure. Teachers then completed six additional questions related to their teaching experience. Specifically, information was obtained on gender, years of teaching experience, years at present school, grade taught, use of the school psychologist, exposure to exceptional children, and satisfaction with school psychological services.

Videotapes.[2] Participants in each condition viewed one, twelve-minute videotaped scenario of consultation interactions between a teacher and a school psychologist. Scenarios depicted condensed versions of three of the four consultation stages outlined by Bergan and Kratochwill (1990), including Problem Identification, Problem Analysis, and Treatment Implementation. The consultation process was condensed to twelve minutes to sample the consultation process and to help ensure teacher participation. The fourth stage of consultation, Treatment Evaluation, was *not* included as the presentation of an intervention outcome would bias viewers who may change their rating based on specific intervention outcome data.

During the first consultation session, the consultant (psychologist) and consultee (teacher) identified the problem and set procedures for preassessment. The target problem in the videotape depicted an elementary school boy who demonstrated disruptive classroom behavior, including talking, arguing, disrupting others, and not completing work. In the second session, problem analysis occurred and a treatment was devised that included procedures for implementation in the classroom and an agenda for monitoring change. The intervention developed for the problem involved a note home program with both parental and teacher reinforcement.

The target problem and intervention were identical across the four conditions. To assure that videotaped scenarios differed only with respect to the manipulated independent variables, participants followed scripts in which the basic content of the sessions described above and verbalizations between consultant and consultee were identical in all four conditions. A basic script was followed for each condition which varied only in the content of the condition. The participants role-playing consultant and consultee had experience in their respective roles. Both had been trained in behavioral consultation and both had been practicing school psychologists. Participants, physical setting, seating, filming, and problem content also were identical in all four scenarios. To assure identical filming, a preset sequence of close-up, individual, and wide-angle shots was followed when filming each scenario. Cues within the scripts assured that nonverbal interactions (e.g., gestures) were also identical.

Prior to being viewed by teachers, videotaped consultation scenarios were also screened by ten school psychology graduate students and ten practicing school psychologists who served as blind raters. These preservice and inservice school psychologists had specific training in behavior modification and behavioral consultation and their ratings were used to assess whether the intended variables (i.e., consultant language and consultee involvement) were adequately represented and could be discriminated. The graduate students completed a twenty-eight-item rating scale assessing the content of each videotape to determine adequate portrayal of technical language, nontechnical language, teacher involvement, and teacher noninvolvement within each condition. The measure included an equal number of items from the consultee and consultant perspectives. The rater was asked to respond to each item on a Likert scale from 1 (extremely so) to 5 (not at all). There were fourteen items representing involvement-noninvolvement (e.g., To what degree was the teacher actively involved in planning the intervention?) and fourteen items representing technical/nontechnical language (e.g., To what degree did the psychologist rely on technical terminology to interpret the child's difficulties to the teacher?). Means computed on the pilot ratings of each videotaped scenario indicated that the independent variables (i.e., language and involvement) were present as intended. The school psychologists were asked to complete a sixteen-item checklist that required discrimination among the four conditions. For each scenario presented, the rater had to check the consultant language (technical or nontechnical) or teacher involvement (involved or noninvolved) condition represented. Results of this analysis indicated that the ten psychologists discriminated the conditions with perfect accuracy.

Independent Variables

Teacher involvement versus teacher noninvolvement. Teachers' preferences toward high versus low degrees of teacher involvement were compared. Some subjects viewed a videotape in which the teacher was highly involved. Other subjects viewed tapes in which teacher involvement was minimal. Under the involvement condition, half the subjects viewed scenarios where the psychologist used technical terms. Other subjects viewed scenarios in which the psychologist used nontechnical language.

In the scenarios depicting teacher involvement, the consultant elicited and utilized the teacher's own ideas to devise a treatment plan and formulate an intervention. In the case of the involvement variable, the consultant, for example, stated during the problem identification interview, "What kinds of information would you be able to gather to help us with this?" The teacher was involved in identifying and analyzing the problem and in designing an intervention. Teachers in both the involvement and noninvolvement scenarios initially offered the same information, but in the teacher involvement condition, teacher input was encouraged and utilized by the psychologist in a collaborative process of analyzing difficulties and coming to some joint conclusions regarding an appropriate intervention. The psychologist in the videotape encouraged the teacher to express her concerns, ideas, observations, and intuitions regarding the problem situation. The psychologist guided the teacher in putting her ideas together and formulating a plan for intervention.

In the contrasting teacher involvement scenarios, a lack of teacher involvement was depicted as the consultant directly stating what should be done and telling the teacher how to do it. In the noninvolvement condition the discussion during problem identification was as follows: "During the next week, keep track of how many times you actually have to speak to Billy during math class for non-instructional reasons." The teacher shared information regarding the problem situation and received all recommendations from the school psychologist regarding an analysis of the difficulties and procedures for intervention. The school psychologist presented her perception and analysis of the problem and then explained the type of intervention that should be used. Thus, the teacher was essentially told what was wrong and what she should do. The school psychologist acted as an expert deciding what the problem was and how the teacher should deal with it.

Technical language versus ordinary language. Subjects viewed taped scenarios involving technical or nontechnical language. Within each condition, subjects viewed scenarios that either did or did not include teacher involvement. One group of subjects viewed a consultant using behavioral terminology to communicate with the teacher. In this scenario, the psychologist/consultant relied on the use of behavioral terms (e.g., reinforcement, contingencies, extinguish, conditioning, shaping, tokens, behavior modification) to appraise the situation and formulate a plan for intervention. For example, in the technical language condition the consultant stated during the problem identification phase: "So, Pam, if you were to give a sequential analysis of these target behaviors, how would you describe them?"

Subjects in the nontechnical language viewed a consultant using non-technical language or the teacher's own terminology to analyze classroom difficulties and formulate a plan. In the nontechnical variation the consultant said: "So, Pam, if you were giving a step-by-step description of the actions you would like changed, how would it go?" Thus, the psychologist in this scenario used nontechnical language (e.g., praise, rewards, stop, change) rather than behavioral terms to make identical appraisals and recommendations of the same problem situation presented by the teacher.

Design. This study utilized a two-by-two factorial design with language and teacher involvement as the manipulated independent variables. The dependent variable, teachers' acceptability ratings of the consultation scenarios, was analyzed along three dimensions to include main effects for language, teacher involvement, and potential interactions.

Measure. After viewing videotaped scenarios, teachers were asked to complete the Intervention Rating Profile-15 (IRF-15), a measure designed to evaluate perceptions of acceptability (Martens et al., 1985). The IRP-15 is a global measure including fifteen items rated on a six-point Likert scale from "strongly disagree" to "strongly agree." Scores can range from fifteen to ninety with higher scores indicating greater acceptability. The IRP-15 is composed of one primary factor, a general acceptability dimension reflecting the degree to which an intervention is judged to be suitable for use in regular classroom settings. The IRP-15 has a good psychometric foundation with a reliability of .98 using Cronbach's alpha (Martens et al., 1985). A factor analysis of the IRP-15 yielded one primary factor with item loadings greater than .82. The brief instructions to the IRP-15 were modified to reflect the purpose of the study, for example, to obtain information that would aid in the selection of styles of consultation and to evaluate consultation which is often provided to a teacher to help children with behavior problems. The only variation to the IRP-15 items made in the present study involved adding and/or substituting the term consultation for/with intervention.

Results

Subjects' acceptability ratings for the consultation scenarios were determined by total scores obtained on the IRP-15. Means and standard deviations for each condition are presented in table 1. A two-way analysis of variance was completed with consultant language and consultee involvement as the independent variables and acceptability via the IRP-15 as the dependent variable. There was no main effect for involvement ($p > .05$) and no main effect for language ($p > .05$).

Table 1. Means and Standard Deviations for Acceptability Ratings

Conditions	Language	
	Technical	Nontechnical
Noninvolved		
Noninvolved M	73.47	58.40
Noninvolved SD	10.80	17.98
Involved		
Involved M	64.20	68.23
Involved SD	18.40	9.49

$n = 15$/cell.

A significant interaction was obtained between consultant language and consultee involvement, $F(1,56) = 6.32, p < .05$. Subjects gave the highest acceptability ratings to the scenario in which teacher involvement was low and the psychologist used technical language. Nontechnical language and low teacher involvement was rated as least acceptable. A Scheffe post hoc test indicated that the difference between the two language groups (Technical and Nontechnical) was relatively greater with the Low Involvement than with the High Involvement, with the net increase in the difference as a function of involvement being about eleven points, with a range (determined from a 95So CI) from 13.06 to 35.14.

Discussion

This study makes two contributions to the study of teacher reactions to behavioral consultation. First, this is the first time consultee involvement has been directly studied via acceptability ratings as a component of the consultation process. Previous researchers (Brigham & Stoerzinger, 1976; Elliott, 1988; Hughes & Falk, 1981; Kazdin, 1980) have suggested the potential importance of consultee involvement but none examined it directly, although our concept is similar to "collaboration" examined in a number of studies. Results of this study indicate that changes in the degree of teacher/consultee involvement or psychologist/consultant directiveness may, in fact, mediate the effect of certain other variables, such as consultant language. Researchers have often suggested that consultants engage in a collaborative relationship with the consultee by emphasizing joint problem solving (Hughes & Falk 1981) and consumer's acceptability of behavioral interventions is enhanced when consultees are given a direct role in implementing and negotiating treatment (Brigham & Stoerzinger, 1976; Kazdin, 1980). The direct involvement of the consultee can also help ensure that an appropriate intervention is planned and that monitoring of the intervention is completed effectively (Clark, 1979; Gutkin & Curtis, 1990). In contrast, Erchul (1987) and Erchul and Chewning (1990) reported results that challenge the traditional concept of collaborative relationships, at least in behavioral consultation. The findings of the present study, that consultees do not show preference for teacher involvement necessarily, may be due to the language used and the measure of acceptability. Of course, researchers have used a variety of different measures across studies, thereby making results more difficult to compare.

Second, an analysis of scores obtained on the IRP-15 indicated that differences in consultant language (technical vs. nontechnical) did not cause acceptability ratings to differ significantly. It therefore appears that although the consultation research has identified consultant language as a variable that may affect consultation acceptability (e.g., Kazdin & Cole, 1981; Witt et al., 1984), a direct relationship may not exist. Researchers have advocated a softening of technical terminology (e.g., Kauffman & Hicente, 1972), but this was many years ago when behavioral techniques were still relatively new and, perhaps, more controversial. Over the years behavioral strategies have been refined, used effectively in the schools, incorporated in parent-training programs, and presented as a component of teacher training programs. There may now be less bias against technical terminology and behavioral techniques. It may now be more acceptable for consultants to achieve a mix of technical and nontechnical language so that the consultee perceives the consultant as an expert with knowledge to share. Unfortunately, in our study we did not obtain information on teacher skill, knowledge, and exposure to behavioral techniques. Future researchers should consider this assessment to extend the findings from our study.

In a study with undergraduate students, presenting behavioral treatments in technical rather than ordinary language was associated with more positive evaluations (Kazdin & Cole, 1981). The authors had not anticipated this result and suggested that the preference for technical language was related to the nature of the subject sample (e.g., college students) or to an increased respectability associated with technical language. In the current study, completed with videotaped scenarios and actual practicing teachers, a similar result was obtained. When the psychologist presented treatments in behavioral language, a more positive evaluation was observed if the teacher was the recipient of advice. Mean scores for acceptability indicated that technical language did receive a more favorable rating when subjects viewed the scenario in which the consultee was not involved. It may be that when the psychologist did not engage the teacher, the consultant became the pivotal point of attention and was, in fact, making a presentation, rather than engaging in a collaborative problem-solving process. If teachers did view the scenarios in which the consultee was not involved from this perspective, the consultant using technical terms may have seemed more knowledgeable, professional, and potentially effective.

In addition to enhancing the credibility and expertise of the consultant, the use of technical language could be beneficial in establishing the credibility of the intervention as a scientific technique. Informal interviews with subjects after the videotapes had been viewed and rated indicated that teachers wanted the psychologist to take a directive role. The implication was that if consultees knew how to handle the difficulty themselves they wouldn't need the consultant in the first place. Teachers also stated that they didn't object to the use of technical terms and that the implication that technical terminology could interfere with effective communication was distasteful in assuming that the teacher had a limited understanding of behavioral terms. This finding is consistent with previous work which indicates that although school psychologists preferred a mental health model of consultation, teachers preferred a behavioral approach largely because the consultant was perceived as specific, competent, efficient, and direct (Medway & Forman, 1980).

With regard to generalizing results from the study, participants viewed and rated only one scenario and were unable to view or rate a contrast between the conditions of language and involvement. Because subjects had not viewed a scenario illustrating the contrasting condition (i.e., consultee noninvolvement), they were less able to rate the adequacy of the condition (i.e., consultee involvement) in the scenario they did view. An alternative strategy in future research would include using two scenarios and have subjects do a comparison rating or separate ratings with controls for ordering effects. Noninvolvement might also be further separated by phase of consultation. For example, the consultant directives regarding what should be done and how to do it might be separated in future research because they are related to different phases of behavioral consultation.

Although the use of a standard measure is desirable, the sole use of the IRP-15 may limit generalizations. The IRP-15 is a good measure of acceptability, but it may be better in future research to use measures designed to assess the effects of language and involvement in addition to the IRP-15.

Finally, generalizability might be limited through the analogue characteristics of this research. For example, the consultation process was condensed to twelve minutes, the consultation was contrived, and teachers viewed a video rather than a real case. However, we believe that understanding of the variables manipulated in this study will be advanced through *both* analogue and naturalistic approaches (see Huebner, 1991, for a similar perspective in special education decision-making research). The tighter controls imposed on analogue research yield greater internal validity, but limit generalizations. Consistency of findings across analogue and naturalistic studies also yields stronger confidence in a research base. Obviously, the cost of naturalistic studies prohibits much needed research in this area.

Notes

Copyright © 1992 by the American Psychological Association. Reproduced, with permission, from Rhoades, M. M., and Kratochwill, T. R. Teacher reactions to behavioral consultation: An analysis of language and involvement. *School Psychology Quarterly*, 1992, 7(1), 47–59. The authors wish to thank Drs. Maribeth Gettinger, Frank Baker, and Steve Elliott for their helpful feedback on the study and participating teachers in the various school districts.

1. Readers might inquire about the number of schools that we had to contact to obtain the sixty subjects. We speculated that there were two reasons for needing this extensive sampling procedure. First, schools required us to submit information to teachers through building principals, who in turn, asked teachers to volunteer. Thus, we had no direct presentation to teachers to stimulate interest. Second, the schools are in an area where requests for subjects from a large research university were frequent and intense.

2. Copies of the videotapes and transcripts are available from the authors for the cost of reproduction and photocopy.

References

Babcock, N. L., & Pryzwansky, W. B. (1983). Models of consultation: Preferences of educational professionals at five stages of service. *Journal of School Psychology, 21,* 359–66.

Bergan, J. R., & Kratochwill, T. R. (1990). *Behavioral consultation in applied settings.* New York: Plenum Press.

Brigham, T. A., & Stoerzinger, A. (1976). An experimental analysis of children's preference for self-selected rewards. In T. A. Brigham, R. P. Hawkins, J. Scott, & T. F. McLaughlin (Eds.), *Behavioral analysis in education: Self-control and reading* (pp. 51–63). Dubuque, IA: Kendall/Hunt.

Clark, R. D. (1979). Should consultants give teachers what they want—a straight answer? *Proceedings of the Eleventh Annual Convention of the National Association of School Psychologists,* p. 3.

Elliott, S. N. (1988). Acceptability of behavioral treatments in educational settings. In J. C. Witt, S. N. Elliott, & F. M. Gresham (Eds.), *The handbook of behavior therapy in education* (pp. 121–50). New York: Plenum.

Erchul, W. P. (1987). A relational communication analysis of control in school consultation. *Professional School Psychology, 2,* 113–24.

Erchul, W. P., & Chewning, T. G. (1990). Behavioral consultation from a request- centered relational communication perspective. *School Psychology Quarterly, 5,* 1–20.

Fairchild, T. N. (1976). School psychological services: An empirical comparison of two models. *Psychology in the Schools, 13,* 156–62.

Fine, M. J., Grantham, V. L., & Wright, J. G. (1979). Personal variables that facilitate or impede consultation. *Psychology in the Schools, 16,* 533–39.

Gutkin, T B., & Curtis, M. (1990). School-based consultation: Theory, techniques, and research. In T. B. Gutkin & C. R. Reynolds (Eds.), *The handbook of school psychology,* 2nd ed., pp. 577–634. New York: Wiley.

Huebner, E. S. (1991). Bias in special education decisions: The contribution of analogue research. *School Psychology Quarterly, 6,* 50–65.

Hughes, J. W., & Falk, R. S. (1981). Resistance, reactance and consultation. *Journal of School Psychology, 10,* 263–68.

Kauffman, J. M., & Hicente, A. P. (1972). Bringing in the Sheaves: Observations on harvesting behavior change in the field. *Journal of School Psychology, 10,* 262–68.

Kazdin, A. E. (1980). Acceptability of time out from reinforcement procedures for disruptive child behavior. *Behavior Therapy, 11,* 329–44.

Kazdin, A. E., & Cole, P. M. (1981). Attitudes and labeling biases toward behavior modification: The effects of labels, content, and jargon. *Behavior Therapy, 12,* 56–68.

Martens, B. K., Peterson, R. L., Witt, J. C., & Cirone, S. (1986). Teacher perceptions of school-based intervention: Ratings of intervention effectiveness, ease of use, and frequency of use. *Exceptional Children, 53,* 213–23.

Martens, B. K., Witt, J. C., Elliott, S. N., & Darveaux, D. K. (1985). Teacher judgments concerning the acceptability of school-based interventions. *Professional Psychology: Research and Practice, 16,* 191–98.

Medway, F. J., & Forman, S. G. (1980). Psychologists' and teachers' reactions to mental health and behavioral school consultation. *Journal of School Psychology, 18,* 338–48.

Reimers, T. M., Wacker, D. P., & Koeppl, G. (1987). Acceptability of behavioral interventions: A review of the literature. *School Psychology Review, 16,* 212–27.

Reinking, R. H., Livesay, G., & Kohl, M. (1978). The effects of consultation style on consultee productivity. *American Journal of Community Psychology, 6,* 283–90.

Wenger, R. D. (1979). Teacher response to collaborative consultation. *Psychology in the Schools, 16,* 127–31.

Wiese, M. R., & Conoley, J. C. (1989). *The relationship of personal problem solving to consultees' preference for collaborative versus expert consultation.* Paper presented at the annual meeting of the American Psychological Association, New Orleans, LA.

Witt, J. C. (1990). Face-to-face verbal interaction in school-based consultation: A review of the literature. *School Psychology Quarterly, 5,* 199–210.

Witt, J. C., & Elliott, S. N. (1985). Acceptability of classroom intervention strategies. In T. R. Kratochwill (Ed.), *Advances in school psychology,* vol. 4, pp. 251–88. Hillsdale, NJ: Erlbaum.

Witt, J. C., & Martens, B. K. (1983). Assessing the acceptability of behavioral interventions used in classrooms. *Psychology in the Schools, 20,* 510–17.

Witt, J. C., Moe, G., Gutkin, T. B., & Andrews, L. (1984). The effect of saying the same thing in different ways: The problem of language and jargon in school-based consultation. *Journal of School Psychology, 22,* 361–67.

Woolfolk, A. E., & Woolfolk, R. L. (1979). Modifying the effect of the behavior modification label. *Behavior Therapy, 10,* 575–78.

Woolfolk, A. E., Woolfolk, R. L., & Wilson, G. T. (1977). A rose by any other name . . . : labeling bias and attitudes toward behavior modification. *Journal of Consulting and Clinical Psychology, 45,* 184–91.

Appendix F: An Example of Mixed Methods Criteria (Principles) for the Validity of Some Holistic Research Designs

Criteria (Standards) of Consistency

A[a]
- Research purposes must be acknowledged.
- Research questions must be acknowledged.
- Consistency must be justified between the question(s) and the purpose(s).
- Consistency must be justified between the purpose(s) and the method(s).
- Method(s) must be justified as capable of fulfilling the purpose(s).
- Method(s) must be justified as capable of addressing the question(s).
- Epistemological assumptions must be consistent among purpose(s), question(s), and methods.
- Methods must be justified as appropriate to these assumptions.

B[b]
- Methods must be epistemologically consistent with findings.
- Findings must be consistent with methods.
- Purposes must be consistent with conclusions.
- Implications must be consistent with purposes.

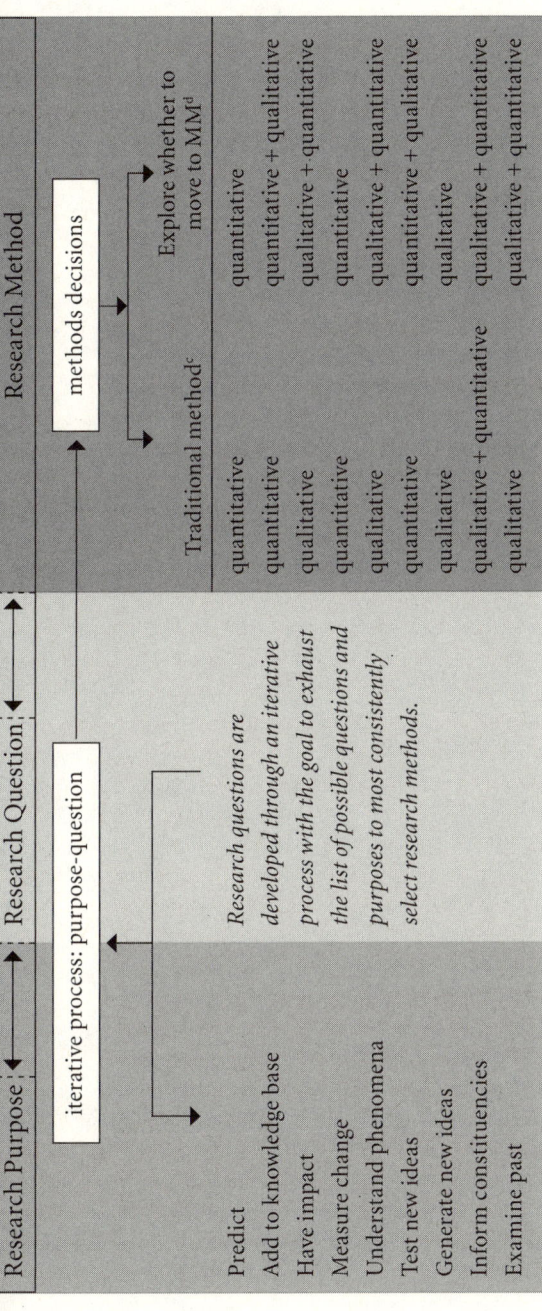

Research Purpose	Research Question	Research Method	
	iterative process: purpose-question	Traditional method[c]	methods decisions — Explore whether to move to MM[d]
Predict	*Research questions are developed through an iterative process with the goal to exhaust the list of possible questions and purposes to most consistently select research methods.*	quantitative	quantitative
Add to knowledge base		quantitative	quantitative + qualitative
Have impact		qualitative	qualitative + quantitative
Measure change		quantitative	quantitative
Understand phenomena		qualitative	qualitative + quantitative
Test new ideas		quantitative	quantitative + qualitative
Generate new ideas		qualitative	qualitative
Inform constituencies		qualitative + quantitative	qualitative + quantitative
Examine past		qualitative	qualitative + quantitative

Source: Examples of research purposes from Newman, Ridenour, Newman, and DeMarco, 2003 (a nine-dimensional typology shown here for illustrative purposes).

Notes: Holistic implies moving beyond limits of a single method. We propose research methods as an interactive continuum rather than a qualitative-quantitative dichotomy, and, so, the term *holistic* is more conceptually accurate in this model of consistency.

[a] The items in *A* are criteria in design.
[b] The items in *B* are criteria in interpretation of results.
[c] Traditionally the paradigm that is consistent with the purpose.
[d] Broadening possible methods beyond the traditional to mixed methods (MM), if multiple purposes.

NOTES

1. Qualitative-Quantitative Research: A False Dichotomy

1. Carolyn S. Ridenour published under the name Carolyn R. Benz prior to 1999.

2. The dramatic impacts of the works of Deming (1991) and Barker (1992) to organizations exemplify strong paradigmatic shifts.

3. *Idealism* refers to the belief that "places special value on ideas and ideals as products of the mind, in comparison with the world as perceived through the senses" (http://www.questia.com/library/philosophy/idealism-philosophy.jsp).

4. *Data* is a term used in both quantitative and qualitative research to refer to the evidence (numerical evidence or narrative evidence). We use it because it is simple in form. We do acknowledge (and use from time to time) the term preferred by Denzin and Lincoln (1994), *empirical materials*, which is more accurate in qualitative research and may be more appealing to some readers.

5. For other discussions of "truth," see, for example, Guba & Lincoln, 1982, 1985; Howe & Eisenhart, 1990; Kvale, 1983; LeCompte & Goetz, 1982; Miles & Huberman, 1984; J. K. Smith, 1983; and Smith & Heshusius, 1986.

6. For our purposes here, Gage's (1963) definition of paradigms is useful "models, patterns. . . . Paradigms are not theories; they are rather ways of thinking or patterns for research that, when carried out, can lead to the development of theory" (p. 95).

2. The Qualitative-Quantitative Research Continuum

1. Campbell and Stanley (1963) wrote the classic treatise on quantitative research design, including the conceptualizations of internal and external validity and their relationships to research design. Their work has probably been and continues to be the most frequently cited work in quantitative behavioral research. They provided the foundation on which other methodologists have built new models (e.g., Krathwohl, 2004). Shadish, Cook, and Campbell (2002) is the first updated version of their model.

2. For a sampling of those researchers who apply multiple methods to their research, see Placek & Dobbs, 1988; Ragin, 1987; Reichardt & Cook, 1979; Shulman, 1986; and Stivers & Srinivasan, 1991.

3. For examples of studies in which the competitive basis of qualitative versus quantitative research is discussed, see Guba, 1978; Guba & Lincoln, 1982, 1989; Howe & Eisenhart, 1990; Kvale, 1983; LeCompte & Goetz, 1982; Miles & Huberman, 1984; J.K. Smith, 1983; and J. K. Smith & Heshusius, 1986.

4. Denzin and Lincoln (2005) chronicled the evolution of qualitative research across the last century, identifying the unique perspectives (epistemologies or methods) that were dominant during seven time periods (using the language "seven moments"). While their chronology places the postpositivist perspective (which we find closest to our own perspective) as emerging and dominant between 1950 and 1986, they acknowledge its continuing legitimacy during the present time. They label the time period of 2000 to the present and beyond, "the methodologically contested present," a time of much tension and conflict among researchers who are opting for a wide variety of "ways of knowing" and qualitative methods. The "methodologically contested present" includes an appreciation for all the dominant perspectives of all the past periods, including, for example, the modernist, postmodernist, postpositivist, critical theorist, and morally driven approaches.

3. Validity and Trustworthiness of Research

1. Many measurement texts provide detailed discussions (e.g., Ary, Jacobs, & Razavieh, 1990; Gay, 1987; McMillan, 2006).

2. Higher-order factorial design is defined in a number of classical statistical textbooks (e.g., Edwards, 1960; Kirk, 1968). Discussion of these designs is beyond the scope of his book.

4. Strategies to Enhance Validity and Trustworthiness

1. Spradley (1979) describes elements of the ethnographic interview, including several types of questions: ethnographic, descriptive, structural, contrast, cultural-ignorance expression, repeating, restating, and so on (p. 67).

5. Applying the Qualitative-Quantitative Interactive Continuum

1. The critique was completed by a spring 1991 University of Akron graduate research class: Sally Gartner, Miriam Keresman, Sandi Sommers, Jayne Speicher, and Brian Tindall.

6. "Science" and a Search for Principles of Practice in Mixed Methods

1. Much of this chapter is taken from "Implementing Mixed Methods Research Designs in the Real World: Purposes, Dilemmas, and New Perspectives," a paper presented in April 2005 by the authors at the annual convention of the American Educational Research Association, Montreal.

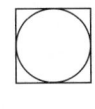

Glossary

accretion measure. "An unobtrusive measure utilizing deposited physical material" (K. D. Bailey, 1978, p. 429; see also Webb et al., 1972).

case record. Condensation of raw data into a manageable, readable package.

case study. An in-depth study of all pertinent aspects of a person, thing, situation, institution, community, and so on (K. D. Bailey, 1978, p. 429; Good, 1963, p. 388; Mouly, 1970, p. 347).

classical theory building versus grounded theory building. In classical theory building, a researcher (1) establishes a concept or proposition, (2) develops hypotheses, and (3) conducts measurements and analyzes to verify the hypotheses. In grounded theory building, a researcher (1) collects and analyzes data, (2) considers only those variables and hypotheses that emerge from the data, and (3) formulates a concept or proposition from the emergent relationships.

closed question. A questionnaire item in which optional response categories are provided for the respondent (K. D. Bailey, 1978, pp. 104, 430; Mouly, 1970, p. 249; Newman, 1976, p. 10).

comparability. The degree to which the ethnographer delineates the constructs generated and the characteristics of the groups studied so that they can be compared to other like and unlike groups (LeCompte and Goetz, 1982, p. 34; Wolcott, 1973).

concurrent validity. The estimate of how well a test correlates with another test that is considered valid.

construct validity. A conglomeration of all other types of validity. (Factor analysis is also frequently used to estimate construct validity.) The degree to which a test has construct validity is related to how well the test estimates the psychological construct being measured. Most frequently, authors refer to construct validity by summarizing several studies that give supportive evidence.

content validity. Also called *logical validity;* an estimate of how representative the test items are of the content or subject matter the test purports to measure. Frequently uses a table of specifications to help estimate the content representativeness.

criterion validity. Name given to predictive and concurrent validity taken together. This type of validity is also called *empirical* or *statistical validity*.

deductive reasoning. Process that is part of the scientific way of knowing, whereby through a series of logical steps, conclusions can be reached based on valid premises; usually considered the basis of quantitative research methods.

descriptive study. A study that uses statistical techniques to describe a sample (for example, by computing the mean) rather than to make inferences from the sample to the population or to use tests of statistical significance (Good, 1963, p. 242; Mouly, 1970, p. 234).

design validity. The extent to which the design is capable of answering the research question and/or the extent to which it can eliminate alternative explanations of the stated relationship (see *internal validity*). If the intent of the study is to generalize, then external validity questions have to be answered to estimate the design validity of the study.

direct-observation. See *participant observation* (Webb et al., 1972, p.113).

epistemology. The study of the nature and grounds of knowledge.

ethnography or **ethnomethodology.** "Studying the commonsense features of everyday life, with emphasis on those things that 'everyone knows'" (K. D. Bailey, 1978, p. 249); social interaction usually is the focus of the study; it comes from traditions in anthropology and sociology (pp. 249–64, 432).

expert-judge validity. Content experts judge whether the test is measuring that which it purports to measure. Similar to *face validity*.

external criticism. The method by which historians determine the genuineness and authenticity of a document (Good, 1963, pp. 200–201).

external validity. The extent to which results of a study are generalizable to other people, groups, settings, or times (Newman & Newman, 1994, p. 229).

face validity. An estimate of participant reaction to the test. If the test appears to the person taking it to be measuring that which it purports to measure, then to that extent it has face validity.

field study. A research strategy in which ethnographic methods are used, and participant-observation strategies are usually employed in natural settings; can be a synonym for *ethnography* (K. D. Bailey, 1978, pp. 221, 433; Lofland, 1971).

focused coding. A part of the grounded theorists' processual analysis, which consists of taking limited sets of codes from the initial coding and applying them to large amounts of data; the level of coding consists of developing categories rather than simply labeling (Charmaz, 1983, p. 116; Glaser & Strauss, 1967).

foundational versus antifoundational assumption. In a *foundational assumption*, reality can be "known" independent of the values of the "knower"; there is certitude and objectivity, as opposed to subjectivity, or "mind-dependent" reality of *antifoundational assumptions* (J. K. Smith, 1990).

grounded theory. A theory generated by or grounded in data rather than being abstract or tentative. Also refers to a research process stressing discovery and theory building through methods of initial and focused coding and memo writing (K. D. Bailey, 1978, p. 44; Charmaz, 1983; Glaser & Strauss, 1967).

indirect observation. A research strategy that is nonreactive, where observational data are collected by unobtrusive means (photographs or films, accretion and/or erosion measures) (K. D. Bailey, 1978, p. 239; Webb et al., 1972).

inductive reasoning. Reasoning from particular facts to a general conclusion; a process that is part of the scientific way of knowing (traditionally used by qualitative researchers) whereby observations or other bits of information (data) are collected, without preconceived notions of their relationships (hypotheses), with the assumption that relationships will become apparent, that conclusions will emerge from the data (Mouly, 1970, p. 30).

initial coding. Part of the grounded theorists' processual analysis, which consists of labeling descriptive information by code names (Charmaz, 1983, p. 113; Glaser & Strauss, 1967).

interactive continuum. A paradigm whereby two phenomena, while conceptually representing bipolar ends of a continuum, can also be used (to a greater or lesser degree in each) at any point along the continuum as the underlying sets of assumptions shift (i.e., interact); it is the descriptor of the research methodology and philosophy proposed in this book.

internal criticism. A method by which historians determine the meaning and trustworthiness of statements within a document (Good, 1963, p. 211).

internal validity. The degree to which all variables, except the one(s) under study, are controlled for in a research design (Keppel, 1973, p. 314; Newman, 1976, p. 231).

interview bias. The effect on the report of interviewer-collected data that comes from the personal attitudes, prejudices, and presuppositions of the interviewer.

interview schedule. A list of questions developed prior to an interview

isomorphism. Literally, "equal in forms." It is used in the philosophy of science to indicate that truth (knowledge) corresponds to reality (J. K. Smith, 1985).

known-group validity. A type of concurrent validity, it is an estimate of how well a test discriminates between identified groups.

measurement validity. The extent to which an instrument measures what it purports to measure.

memo writing. Part of the grounded theorists' processual analysis between coding and writing the results by which the researchers elaborate on the categories of data and the relationships among them. Memos are subsequently "sorted" and "integrated." Some researchers write many short memos on many categorical relationships, and over time, the analytical level (of the ideas and the memos) becomes more abstract, and through this, theory is built (Charmaz, 1983, p. 120; Glaser & Strauss, 1967).

multiple regression analysis. Sometimes referred to as the general case of *least squares solution,* the basic underlying concept is that more than one predictor variable is used to predict one criterion variable; sometimes interchanged with the concept of *multiple correlation technique.*

multivariate. A condition in which there are two or more dependent (or criterion) variables being predicted by two or more independent (predictor) variables.

naturalistic inquiry. The type of research that is generally subsumed under qualitative methods; it is based on the underlying assumptions that knowledge about reality is mind dependent not value free, that hypotheses are always working hypotheses, and that reality is not a single construct but multiple constructs. Techniques such as interview, observation, unobtrusive measurement, and document analysis are typically used to glean information in natural settings.

nonparticipant observation. A research situation in which the researcher collects data as an outsider and does not participate in ongoing activities (K. D. Bailey, 1978, p. 215).

nonstructured interview. A research strategy whereby the interviewer has a topic in mind but no predetermined questions (K. D. Bailey, 1978, p. 176; Newman, 1976, p. 12).

open-ended question. A type of questionnaire item for which predetermined response options are not provided (K. D. Bailey, 1978, pp. 104, 435; Mouly, 1970, p. 249; Newman, 1976, p. 9).

partially structured interview. A research strategy whereby some questions are predetermined and the interviewer also uses open-ended questions and probes to explore more in-depth reasons for answers.

participant observation. A research situation in which the researcher is a regular participant in the activities being observed while he/she collects data; the dual role is usually not known to the other participants (K. D. Bailey, 1978, pp. 215, 435; Webb et al.: 1972, p. 115). See also *direct observation.*

phenomenological research. A qualitative research method founded by Edmund Husserl as a reaction against the empiricist conception of the world as an "objective universe of facts" (Kvale, 1983). The researcher is involved in three aspects: "open description," "investigation of essences," and "phenomenological reduction," whereby interview responses are recorded, transcribed, and reviewed for central themes of meaning (Mitchell, 1990).

predictive validity. The estimate of how well a test predicts eventual outcome.

primary source. A document or data set provided by actual witnesses to the incident or phenomenon under study (Good, 1963, p. 194; Mouly, 1970, p. 213).

processual analysis. A research method used by grounded theorists in which the process of coding the data constitutes the analysis of the data (Charmaz, 1983, p, 117; Glaser & Strauss, 1967).

qualitative analysis. Data analysis in which the aim is building an explanation of meaning of persons' lives from their perspectives. It is generally inductive in approach, is based originally on the naturalistic assumption that reality is mind dependent (i.e., can only be known as it is interpreted and has "meaning" for the observer), is usually of single-subject design, and generally deals with nominal data (K. D. Bailey, 1978, p. 436; LeCompte & Goetz, 1982; Reichardt & Cook, 1.979, p. 10; Rist, 1977; J. K. Smith, 1983).

quantitative analysis. Data analysis in which the aim is theory testing. It is generally deductive in approach; is based originally on the rationalistic assumption that reality is mind independent (i.e., can separate the observer from the object of study), and has as its goal generalizability. It usually deals with ordinal, interval, or ratio data (K. D. Bailey, 1978, p. 436; LeCompte & Goetz, 1982; Reichardt & Cook, 1979, p. 10; Rist, 1977; J. K. Smith: 1983).

rationalist versus naturalist philosophy. *Rationalist philosophy* holds that one can "know" reality as objective phenomena, outside the influence of the "knower" and his/her values; whereas *naturalist philosophy* holds that one cannot separate the "known" reality from the values of the "knower"; in other words, knowledge that is mind independent versus knowledge that is mind dependent (Mouly, 1970; J. K, Smith, 1983, 1985).

raw case data. All information (person, place, or thing) collected about a research topic.

reliability. A value indicating the internal consistency of a measure or the repeatability of a measure or finding; the extent to which a result or measurement will be the same value every time it is measured (Keppel, 1973, p. 310; Newman & Newman, 1994).

scientific method. The step-by-step process by which theory is both generated and verified; both an inductive and a deductive process. Generally a phenomenon is observed, set(s) of relationships is (are) stated, a hypothesis is stated, a design is created to test the hypothesis, data are collected, data are analyzed, the results are concluded, the hypothesis is verified or refuted, and theory is refined (Good, 1963, p. 5).

secondary source. A source containing an intermediate person's (not the actual witnesses') reporting of the event or phenomenon under study (Good, 1963, p. 194; Mouly, 1970, p. 213).

sorting memos. Part of the grounded theorists' processual analysis. It follows initial coding, focused coding, and writing memos; the researcher analyzes relationships among memos and pulls them together to develop theoretical relationships (Charmaz, 1983).

structured interview. A research strategy in which the interviewer reads each question and the possible answers; the respondent responds; and the response is recorded (K. D. Bailey, 1978, p. 170; Newman, 1976, p. 12).

structured observation. A research situation in which data are collected by the observer's use of predetermined categories of behaviors (K. D. Bailey, 1978, pp. 216, 231).

synthesis. The process of blending external criticism and internal criticism to report historical data accurately (Mouly, 1970, p. 211).

theoretical sampling. The sampling of additional data to develop an emerging theory (Charmaz, 1983 p. 124–25). "The process of data collection for generating theory whereby the analyst jointly collects, codes, and analyzes his data and decides what data to collect next and where to find them, in order to develop his theory as it emerges" (Glaser & Strauss, 1967, p, 45).

theory building versus theory testing. The difference some researchers use to dichotomize qualitative and quantitative philosophies and methods; in *theory building*, qualitative methodologists collect data with neither a theoretical base nor a hypothesis and use the data to generate categories as well as statements of relationships. In *theory testing*, quantitative investigators begin with a theoretically based hypothesis and collect previously established categories of data to test the viability of that hypothesis. Some qualitative researchers use the word *theory* to refer to the researcher's point of view or lens on what is being studied.

translatability. The degree to which the research methods, analytical categories, and characteristics of phenomena and groups are explicitly described by ethnographers so that comparisons with other groups can be made (LeCompte & Goetz, 1982, p. 34).

triangulation. The combining of two or more data-collection methods and/or data sources into one design (K. D. Bailey, 1978, p. 239; Jick, 1979; LeCompte & Goetz, 1982, p. 35; Webb et al., 1972).

univariate. One variable (dependent or criterion variable) is being predicted by another set of variables (independent or predictor variable). This set of predictor variables could include one or more than one variable.

unobtrusive measure. A nonreactive measure in which the behavior of the participants being studied is not changed because they do not know research is being conducted (K. D. Bailey, 1978, p. 239; Webb et al., 1972).

unstructured observation. A research strategy in which observational data are collected without predetermined categories to look for or hypotheses to guide the observation (K. D. Bailey, 1978, p. 216).

validity. In the context of quantitative measurement or instrumentation, the degree to which one actually is measuring what one wishes to measure; several types exist (Keppel, 1973, p. 310; Mouly, 1970, p. 118; Newman, 1976, pp. 56, 240). In the context of qualitative research, the extent to which the results are trustworthy. Trustworthiness replaces validity in qualitative research. Guba and Lincoln replace validity with several notions of "goodness criteria" in qualitative research (see Guba & Lincoln, 2005).

verstehen. The researcher attempts to portray the meaning of the lives of those he or she studies from their (those being studied) point of view.

REFERENCES

Alexander, J. C., & Harman, R. L. (1988). One counselor's intervention in the aftermath of a middle school student's suicide: A case study. *Journal of Counseling and Development, 66,* 283–85.

Angrosino, M. V. (2005). Recontextualizing observation: Ethnography, pedagogy, and the prospects for a progressive political agenda. In N. K. Denzin & Y. S. Lincoln (Eds.), *The Sage handbook of qualitative research* (3rd ed., pp. 729–45). Thousand Oaks, CA: Sage.

Ary, D., Jacobs, L. C., & Razavieh, A. (1990). *Introduction to research in education* (4th ed.). Fort Worth, TX: Holt, Rinehart & Winston.

Bailey, K. D. (1978). *Methods of social research.* New York: Macmillan.

Barker, J. A. (1992). *Future edge: Discovering the new paradigms of success.* New York: Morrow.

Bar-On, R., & Parker, J. D. A. (Eds.). (2000). *The handbook of emotional intelligence: Theory development, assessment, and application at home, school, and in the workplace.* San Francisco: Jossey-Bass.

Bauer, H. H. (1992). *Scientific literacy and the myth of the scientific method.* Chicago: University of Chicago Press.

Becker, H. S., & Geer, B. (1960). Participant observation: The analysis of qualitative field data. In R. N. Adams & J. J. Preiss (Eds.), *Human organizational research: Field relations and techniques* (pp. 267–89). Homewood, IL: Dorsey.

Benz, C., & Newman, I. (1986). *Qualitative-quantitative interactive continuum: A model and application to teacher education evaluation.* Paper presented at the American Association of Colleges for Teacher Education meeting, Chicago, IL (ERIC Document Reproduction Service No. ED 269406).

Blumer, H. (1980). Comment, Mead and Blumer: The convergent methodological perspective of social behaviorism and symbolic interactionism. *American Sociological Review, 45,* 409–19.

Boostrom, R. (1994). Learning to pay attention. *International Journal of Qualitative Studies in Education, 7*(1), 51–64.

Brewer, J., & Hunter, A. (1989). *Multimethod research: A synthesis of styles.* Newbury Park, CA: Sage.

Campbell, D. T., & Fiske, D. W. (1959). Convergent and discriminant validation by the multitrait-multimethod matrix. *Psychological Bulletin, 56,* 81–105.

Campbell, D. T., & Stanley, J. C. (1963). *Experimental and quasi-experimental designs for research.* Chicago: Rand McNally.

Charmaz, K. (1983). The grounded theory method: An explication and interpretation. In R. M. Emerson (Ed.), *Contemporary field research: A collection of readings* (pp. 109–26). Boston: Little, Brown.

———. (2000). Grounded theory: Objectivist and constructivist methods. In N. K. Denzin & Y. S. Lincoln (Eds.), *The Sage handbook of qualitative research* (2nd ed., pp. 509–35). Thousand Oaks, CA: Sage.

———. (2005). Grounded theory in the twenty-first century: Applications for advancing social justice studies. In N. K. Denzin & Y. S. Lincoln (Eds.), *The Sage handbook of qualitative research* (3rd ed., pp. 507–35). Thousand Oaks, CA: Sage.

Comte, A. (1974). *Discours sur l'esprit positif.* Paris: Librairie Philosophique. (Original work published 1844).

Creswell, J. W. (2005). *Educational research: Planning, conducting, and evaluating quantitative and qualitative research* (2nd ed.). Upper Saddle River, NJ: Pearson Education.

Cronbach, L., & Meehl, P. E. (1955). Construct validity in psychological tests. *Psychological Bulletin, 52*(4), 281–302.

Culbertson, J. A. (1988). A century's quest for a knowledge base. In N. J. Boyan (Ed.), *Handbook of research on educational administration* (pp. 3–26). New York: Longman.

Curtis, J. M. (1981). Effect of therapist's self-disclosure on patients' impressions of empathy, competence, and trust in an analogue of a psychotherapeutic interaction. *Psychological Reports, 48,* 127–36.

Deming, W. E. (1991). *Out of the crisis.* Cambridge, MA: Massachusetts Institute of Technology, Center for Advanced Engineering Study.

Denzin, N. K. (1978). The logic of naturalistic inquiry. In N. K. Denzin (Ed.), *Sociological methods: A sourcebook* (2nd ed.). New York: McGraw-Hill.

———. (1988). *The research act* (Rev. ed.). New York: McGraw-Hill.

———. (1994). Evaluating qualitative research in the poststructural moment: The lessons James Joyce teaches us. *International Journal of Qualitative Studies in Education, 7*(4), 295–308.

Denzin, N. K., & Lincoln, Y. S. (Eds.). (1994). *Handbook of qualitative research.* Thousand Oaks, CA: Sage.

——— (Eds.). (2000). *Handbook of qualitative research* (2nd ed.). Thousand Oaks, CA: Sage.

——— (Eds.). (2005). *The Sage handbook of qualitative research* (3rd ed.). Thousand Oaks, CA: Sage.

Dewey, J. (1929). *The sources of a science of education.* New York: Liveright.

Diesing, P. (1991). *How does social science work? Reflections on practice.* Pittsburgh, PA: University of Pittsburgh Press.

Donmoyer, R. (1990). Generalizability and the single-case study. In E. Eisner & A. Peshkin (Eds.), *Qualitative inquiry in education: The continuing debate* (pp. 175–200). New York: Teachers College Press.

Edwards, A. L. (1960). *Experimental design in psychological research.* New York: Rinehart.

Egan, K. (2005). Students' development in theory and practice: The doubtful role of research. *Harvard Educational Review, 75*(1), 25–41.

Eisner, E. W., & Peshkin, A. (Eds.). (1990). *Qualitative inquiry in education: The continuing debate.* New York: Teachers College Press.

Fetterman, D. M. (1989). *Ethnography: Step by step.* Newbury Park, CA: Sage.

Fink, A. S. (2000). The role of the researcher in the qualitative research progress. A potential barrier to archiving qualitative data. *Forum Qualitative Sozialforschung [Forum Qualitative Social Research], 1*(3). Available: http://www.qualitative-research.net/fqs-texte/3-00/3-00fink-e.htm.

Foerstner, S. B., Newman, I., & Koenig, D. (1985, October). *To laugh or not to laugh: Why? Its measurement and meaning.* Paper presented at the meeting of the Midwestern Educational Research Association, Chicago, IL.

Fontana, A., & Frey, J. H. (2005). The interview: From neutral stance to political involvement. In N. K. Denzin & Y. S. Lincoln (Eds.), *The Sage handbook of qualitative research* (3rd ed., pp. 695–727). Thousand Oaks, CA: Sage.

Frechtling, J., & Sharp, L. (Eds.). (1997). *User-friendly handbook for mixed method evaluations.* Arlington, VA: Directorate for Education and Human Resources, National Science Foundation, Division of Research, Evaluation, and Communication.

Fuller, M. L. (1994). The monocultural graduate in the multicultural environment: A challenge for teacher educators. *Journal of Teacher Education, 45*(4), 269–77. Available from http://search.ebscohost.com/login.aspx?direct=true&db=ehh&AN=9411081903&site=ehost-live.

Gage, N. L. (1963). *Handbook of research on teaching.* Chicago: Rand McNally.

Gall, J. P., Gall, M. D., & Borg, W. R. (2005). *Applying educational research: A practical guide*. Boston: Pearson.

Gay, L. R. (1987). *Educational research: Competencies for analysis and application* (3rd ed.). Columbus, OH: Merrill.

Geertz, C. (1973). Thick description: Toward an interpretive theory of culture. In *The interpretation of cultures* (pp. 3–30). New York: Basic.

Glaser, B. G., & Strauss, A. L. (1967). *The discovery of grounded theory: Strategies for qualitative research*. Chicago: Aldine.

Goetz, J., & LeCompte, M. (1984). *Ethnography and qualitative design in educational research*. Orlando, FL: Academic.

Good, C. V. (1963). *Introduction to educational research: Methodology of design in the behavioral and social sciences*. New York: Appleton-Century-Crofts.

Greene, J. C., & Caracelli, V. J. (Eds.). (1997). *Advances in mixed-method evaluation: The challenges and benefits of integrating diverse paradigms*. San Francisco: Jossey-Bass.

———. (2003). Making paradigmatic sense of mixed methods practice. In A. Tashakkori & C. Teddlie (Eds.), *Handbook of mixed methods in social and behavioral research* (pp. 91–110). Thousand Oaks, CA: Sage.

Guba, E. G. (1978). *Toward a methodology of naturalistic inquiry in educational evaluation*. (CSE Monograph Series in Evaluation No. 8). Los Angeles: University of California, Center for Study of Evaluation.

Guba, E. G., & Lincoln, Y. S. (1982). Epistemological and methodological bases of naturalistic inquiry. *Educational Communications and Technology Journal, 30*(4), 233–52.

———. (1985, October). *The countenances of fourth generation evaluation: Description, judgment and negotiation*. Paper presented at the meeting of the Evaluation Network, Toronto, Ontario, Canada.

———. (2005). Paradigmatic controversies, contradictions, and emerging confluences. In N. K. Denzin & Y. S. Lincoln (Eds.), *The Sage handbook of qualitative research* (3rd ed., pp. 191–215). Thousand Oaks, CA: Sage.

Gubrium, J. F., & Holstein, J. A. (2002). *Handbook of interview research: Context and method*. Thousand Oaks, CA: Sage.

Hakim, C. (1987). *Research design: Strategies and choices in the design of social research*. Boston: Allen & Unwin.

Hammersley, M. (1992). Some reflections on ethnography and validity. *International Journal of Qualitative Studies in Education, 5*(3), 195–203.

Howard, D. C. P. (1994). Human-computer interactions: A phenomenological examination of the adult first-time computer experience. *International Journal of Qualitative Studies in Education* 7(1), 33–50.

Howe, K. R. (1985). Two dogmas of educational research. *Educational Researcher,* 14(8), 10–18.

———. (1988). Against the quantitative-qualitative incompatibility thesis; or, Dogmas die hard. *Educational Researcher,* 17(8), 10–16.

Howe, K., & Eisenhart, M. (1990). Standards for qualitative (and quantitative) research: A prolegomenon. *Educational Researcher* 19(4), 2–9.

Jick, T. D. (1979). Mixing qualitative and quantitative methods: Triangulation in action. *Administrative Science Quarterly,* 24, 602–11.

Johnson, B., & Christensen, L. (2004). *Educational research: Quantitative, qualitative, and mixed approaches.* Boston: Pearson.

Johnson, B., & Turner, L. A. (2003). Data collection strategies in mixed methods research. In A. Tashakkori & C. Teddlie (Eds.), *Handbook of mixed methods in social and behavioral research* (pp.297–320). Thousand Oaks, CA: Sage.

Keppel, G. (1973). *Design and analysis: A researcher's handbook.* Englewood Cliffs, NJ: Prentice-Hall.

Kerlinger, F. N., & Lee, H. B. (2000). *Foundations of behavioral research* (4th ed.). Fort Worth, TX: Harcourt College.

Kirk, R. E. (1968). *Experimental design: Procedures for the behavioral sciences.* Belmont, CA: Brooks.

Kitzinger, C. (1987). *The social construction of lesbianism.* Newbury Park, CA: Sage.

Krathwohl, D. (2004). *Methods of educational and social science research: An integrated approach* (2nd ed.). Long Grove, IL: Waveland.

Kuhn, T. S. (1962). *The structure of scientific revolutions.* Chicago: University of Chicago Press.

———. (1970). *The structure of scientific revolutions* (2nd ed.). Chicago: University of Chicago Press.

Kvale, S. (1983). The qualitative research interview: A phenomenological and a hermeneutical mode of understanding. *Journal of Phenomenological Psychology,* 14(2), 171–95.

———. (1995). The social construction of validity. *Qualitative Inquiry,* 1(1), 19–40.

———. (1996). *Interviews: An introduction to qualitative research interviewing.* Thousand Oaks, CA: Sage.

Lagemann, E. C. (2000). *An elusive science: The troubling history of education research.* Chicago: University of Chicago Press.

Lather, P. (1986). Issues of validity in openly ideological research. *Interchange, 17*(4), 63–84.

———. (1995). The validity of angels: Interpretive and textual strategies in researching the lives of women with HIV/AIDS. *Qualitative Inquiry, 1*(1), 41–68.

LeCompte, M. D., & Goetz, J. P. (1982). Problems of reliability and validity in ethnographic research. *Review of Educational Research, 52*(1), 31–60.

Lincoln, Y. S. (1990). The making of a constructivist: Remembrance of transformations past. In E. Guba (Ed.), *The paradigm dialog* (pp. 67–87). Newbury Park, CA: Sage.

———. (1990). Toward a categorical imperative for qualitative research. In E. W. Eisner & A. Peshkin (Eds.), *Qualitative inquiry in education: The continuing debate* (pp. 277–95). New York: Teachers College Press.

———. (1995). Emerging criteria for quality in qualitative research and interpretative research. *Qualitative Inquiry, 1*(3), 275–89.

Lincoln, Y. S., & Guba, E. G. (1985). *Naturalistic inquiry.* Beverly Hills, CA: Sage.

———. (2000). Paradigmatic controversies, contradictions, and emerging confluences. In N. K. Denzin & Y. S. Lincoln (Eds.), *Handbook of qualitative research* (2nd ed., pp. 163–88). Thousand Oaks, CA: Sage.

Linn, R. L., & Miller, M. D. (2005). *Measurement and assessment in teaching* (9th ed.). Upper Saddle River, NJ: Pearson Education.

Lofland, J. (1971). *Analyzing social settings. A guide to qualitative observation and analysis.* Belmont, CA: Wadsworth.

Lombard, G. J. (1991, April). *Closing the loop: Adding the qualitative dimension to critical incidents.* Paper presented at the meeting of the American Educational Research Association, Chicago, IL.

Lyotard, J. E. (1984). *The postmodern condition: A report on knowledge.* Manchester, UK: Manchester University Press.

MacCorquodale, K., & Meehl, P. E. (1948). On a distinction between hypothetical constructs and intervening variables. *Psychological Review, 55,* 95–107.

McMillan, J. H. (2006). *Classroom assessment: Principles and practice for effective standards-based instruction* (4th ed.). New York: Allyn & Bacon.

McNeil, K., Newman, I., & Kelly, F. J. (1996). *Testing research hypotheses with the general linear model.* Carbondale: Southern Illinois University Press.

Medawar, P. B. (1984). *The limits of science.* New York: Harper & Row.
Merriam, S. B. (1988). *Case study research in education: A qualitative approach.* San Francisco: Jossey-Bass.
Mertens, D. M. (2003). Mixed methods and the politics of human research: The transformative-emancipatory perspective. In A. Tashakkori & C. Teddlie (Eds.), *Handbook of mixed methods in social and behavioral research* (pp. 135–64). Thousand Oaks, CA: Sage.
Miles, M. B., & Huberman, A. M. (1984). Drawing valid meaning from qualitative data: Toward a shared craft. *Educational Researcher, 13*(5), 20–30.
Miller, L., & Lieberman, A. (1988). School improvement in the United States: Nuance and numbers. *Qualitative Studies in Education, 1*(1), 3–19.
Mitchell, J. G. (1990). *Re-visioning educational leadership: A phenomenological approach.* New York: Garland.
Morse, J. M. (2003). Principles of mixed methods and multimethod research design. In A. Tashakkori & C. Teddlie (Eds.), *Handbook of mixed methods in social and behavioral research* (pp. 189–208). Thousand Oaks, CA: Sage.
Mouly, G. J. (1970). *The science of educational research* (2nd ed.). New York: Van Nostrand Reinhold.
Moustaskas, C. (1994). *Phenomenological research methods.* Thousand Oaks, CA: Sage.
Newman, I. (1976). *Basic procedures in conducting survey research.* Akron, OH: University of Akron.
Newman, I., & Benz, C. R. (1998). *Qualitative-quantitative research methodology: Exploring the interactive continuum.* Carbondale: Southern Illinois University Press.
Newman, I., & McNeil, K. (1998). *Conducting survey research in the social sciences.* Lanham, MD: University Press of America.
Newman, I., McNeil, K., & Fraas, J. (2004). Two methods of estimating a study's replicability. *Mid-Western Educational Researcher, 17*(2), 36–40.
Newman, I., & Newman, C. (1994). *Conceptual statistics for beginners* (2nd ed.). Lanham, MD: University Press of America.
Newman, I., Newman, C., Brown, R., & McNeely, S. (2006). *Conceptual statistics for beginners* (3rd ed.). Lanham, MD: University Press of America.
Newman, I., Ridenour, C. S., Newman, C., & DeMarco, G. M. P., Jr. (2003). Typology of purpose. In A. Tashakkori & C. Teddlie (Eds.), *Handbook of mixed methods in social and behavioral research* (pp.167–88). Thousand Oaks, CA: Sage.

Norell, M. (2005, May 13). Interview: New exhibit updates story of dinosaur era. In *Morning Edition*. Washington, DC: National Public Radio.

Ödman, P. J. (1992). Interpreting the past. *International Journal of Qualitative Studies in Education, 5*(2), 167–84.

Olson, L. (1986, January 15). "Positive science" or "normative principles"? *Education Week*. Retrieved from http://www.edweek.org/ew/articles/1986/01/15/06160043.h05.html.

Onwuegbuzie, A. J., & Daniel, L. G. (2003, February 19). Typology of analytical and interpretational errors in quantitative and qualitative educational research. *Current Issues in Education* [On-line], *6*(2). Available: http://cie.ed.asu.edu/volume6/number2/.

Page, R. N. (1997). Teaching about validity. *International Journal of Qualitative Studies in Education, 10*(2), 145–55.

Patton, M. Q. (1980). *Qualitative evaluation methods*. Beverly Hills, CA: Sage.

———. (1990). *Qualitative evaluation and research methods* (2nd ed.). Newbury Park, CA: Sage.

———. (2002). *Qualitative research and evaluation methods* (3rd ed.). Thousand Oaks, CA: Sage.

Phillips, D. C., & Barbules, N. C. (2000). *Postpositivism and educational research*. Lanham, MD: Rowman & Littlefield.

Placek, J. H., & Dobbs, P. (1988). A critical incident study of preservice teachers' beliefs about teaching success and nonsuccess. *Research Quarterly for Exercise and Sport, 59*(4), 351–58.

Polkinghorne, D. E. (1983). *Methodology for the human sciences: Systems of inquiry*. Albany: State University of New York Press.

———. (1991, April). *Generalization and qualitative research: Issues of external validity*. Paper presented at the meeting of the American Educational Research Association, Chicago, IL.

Popper, K. R. (1959). *The logic of scientific discovery*. New York: Basic.

———. (1962). *Conjectures and refutations: The growth of scientific knowledge*. New York: Basic.

Ragin, C. C. (1987). *The comparative methods: Moving beyond qualitative and quantitative strategies*. Berkeley: University of California Press.

Raudenbush, S. W., & Bryk, A. S. (2002). *Hierarchical linear models: Applications and data analysis methods*. Thousand Oaks, CA: Sage.

Reichardt, C. S., & Cook, T. D. (1979). Beyond qualitative versus quantitative methods. In T. D. Cook & C. S. Reichardt (Eds.), *Qualitative and quantitative methods in evaluation research* (pp. 7–32). Beverly Hills, CA: Sage.

Reichardt, C. S., & Rallis, S. F. (1994). *The qualitative-quantitative debate: New perspectives.* San Francisco: Jossey-Bass.

Rhoades, M. M., & Kratochwill, T. R. (1992). Teacher reactions to behavioral consultation: An analysis of language and involvement. *School Psychology Quarterly, 7*(1), 47–59.

Ricoeur, P. (1988). *From text to action: An anthology on hermeneutics.* Stockholm, Sweden: Symposium.

Ridenour, C. S., & Newman, I. (2004, October). *"Themes are not variables" and mixed methods are not a "panacea" for educational researchers.* Paper presented at the annual meeting of the Midwestern Educational Research Association, Columbus, OH.

———. (2005, April). *Implementing mixed methods research designs in the real world: Purposes, dilemmas, and new perspectives.* Paper presented at the annual meeting of the American Educational Research Association, Montreal, Quebec, Canada.

Ridenour, C. S., Newman, I., De Marco, G. M. P., & Newman, C. (2003, October). *The perils of neglecting "purpose" in education research and strategies for rethinking its crucial place in design: An interaction session.* Paper presented at the Midwestern Educational Research Association, Columbus, OH.

Rosenthal, R., & Rosnow, R. L. (1991). *Essentials of behavioral research: Methods of data analysis* (2nd ed.). New York: McGraw-Hill.

Sandelowksi, M. (2003). Tables or tableaux? The challenges of writing and reading mixed methods studies. In A. Tashakkori & C. Teddlie (Eds.), *Handbook of mixed methods in social and behavioral research* (pp. 321–50). Thousand Oaks, CA: Sage.

Schatzman, L., & Strauss, A. L. (1973). *Field research: Strategies for a natural sociology.* Englewood Cliffs, NJ: Prentice-Hall.

Shadish, W. R., Cook, T. D., & Campbell, D. T. (2002). *Experimental and quasi-experimental designs for generalized causal inference.* Boston: Houghton Mifflin.

Shafer, R. J. (Ed.). (1974). *A guide to historical method.* Homewood, IL: Dorsey.

Shulman, L. S. (1986). Paradigms and research programs in the study of teaching: A contemporary perspective. In M. C. Wittrock (Ed.), *Handbook of research on teaching* (3rd ed., pp. 3–36). New York: Macmillan.

———. (1987). Knowledge and teaching: Foundations of the new reform. *Harvard Educational Review 57*(1), 1–22.

Sindell, P. S. (1969). Anthropological approaches to the study of education. *Review of Educational Research, 39,* 593–605.

Smith, J. K. (1983). Quantitative versus qualitative research: An attempt to clarify the issue. *Educational Researcher, 12*(3), 6–13.

———. (1985, April). *Closing down the conversation: The end of the qualitative-quantitative debates in educational inquiry.* Paper presented at the meeting of the American Educational Research Association, Chicago, IL.

———. (1990). Alternative research paradigms and the problem of criteria. In E. Guba (Ed.), *The paradigm dialog* (pp.167–87). Newbury Park, CA: Sage.

Smith, J. K., & Heshusius, L. (1986). Closing down the conversation: The end of the quantitative-qualitative debate among educational inquires. *Educational Researcher, 15*(1), 4–12.

Smith, L. M. (1967). The microethnography of the classroom. *Psychology in the Schools, 4,* 216–21.

Spencer, H. (1910). *Essays: Scientific, political, and speculative.* New York: Appleton.

Spicer, N. (2005).*Using mixed methods in social research: An introduction for students and researchers.* Thousands Oak, CA: Sage.

Spindler, G. D. (Ed.). (1974). *Educational and cultural process: Toward an anthropology of education.* New York: Holt, Rinehart, & Winston.

Spradley, J. P. (1979). *The ethnographic interview.* New York: Holt, Rinehart, & Winston.

Stake, R. E. (1981). Case study methodology: An epistemological advocacy. In W. W. Welch (Ed.), *Case study methodology in educational evaluation: Proceedings of the 1981 Minnesota Evaluation Conference* (pp. 31–40). Minneapolis: Minnesota Research and Evaluation Center.

Stivers, E., & Srinivasan, R. (1991, April). *Using mathematics in qualitative research.* Paper presented at the meeting of the American Educational Research Association, Chicago, IL.

Tashakkori, A., & Teddlie, C. (1998). *Mixed methodology: Combining qualitative and quantitative approaches.* Thousand Oaks, CA: Sage.

———. (2003). The past and future of mixed methods research: From data triangulation to mixed model designs. In A. Tashakkori & C. Teddlie (Eds.), *Handbook of mixed methods in social and behavioral research* (pp. 671–701). Thousand Oaks, CA: Sage.

Tierney, W. G. (1993). The cedar closet. *International Journal of Qualitative Studies in Education, 6*(4), 303–14.

Toma, J. D. (2006). Approaching rigor in applied qualitative research. In C. F. Conrad & R. C. Serlin (Eds.), *The Sage handbook for research in education: Engaging ideas and enriching inquiry*, pp. 405–23). Thousand Oaks, CA: Sage.

Trankell, A. (1972). *Reliability of evidence.* Stockholm, Sweden: Beckmans.

Trueba, H. T. (1991, April). Back to basics in educational research: The uses and abuses of "culture" and "ethnography" in educational research. Paper presented at the meeting of the American Educational Research Association, Chicago, IL.

Trueba, H. T., Jacobs, L., & Kirton, E. (1990). *Cultural conflict and adaptation: The case of Hmong children in American society.* New York: Falmer.

Van Manen, M. (1990). *Researching lived experience. Human science for an action sensitive pedagogy.* London, Ontario, Canada: Althouse.

Vidich, A. J., & Lyman, S. M. (1994). Qualitative methods: Their history in sociology and anthropology. In N. K. Denzin & Y. S. Lincoln (Eds.), *Handbook of qualitative research* (pp. 23–59). Thousand Oaks, CA: Sage.

Wax, M. L., Diamond, S., & Gearing, F O. (Eds.). (1971). *Anthropological perspectives on education.* New York: Basic.

Webb, E. J., Campbell, D. T., Schwartz, R. D., & Sechrest, L. (1972). *Unobtrusive measures: Nonreactive research in the social sciences.* Chicago: Rand McNally.

Wertz, F J. (1986). The question of reliability of psychological research. *Journal of Phenomenological Psychology, 17,* 81–205.

Wiersma, W. (1980). *Research methods in education: An introduction* (3rd ed.). Itasca, IL: Peacock.

Winter, G. (2000, March). A comparative discussion of the notion of "validity" in qualitative and quantitative research. [58 paragraphs]. *The Qualitative Report* [On-line serial], *4* (3/4). Available: http://www.nova.edu/ssss/QR/QR4-3/winter.html.

Wolcott, H. F (1973). *The man in the principal's office: An ethnography.* New York: Holt, Rinehart, & Winston.

———. (1990). On seeking—and rejecting—validity in qualitative research. In E. W. Eisner & A. Peshkin (Eds.), *Qualitative inquiry in education: The continuing debate* (pp. 121–52). New York: Teachers College Press.

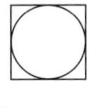

INDEX

Page locators in italics refer to figures.

action validity, 55, 56
active variable, 48
aggregate generalizability, 61
Alexander, Jo Ann C., critique of study by, 97–100, 123–29
alpha levels, 41–42
American Educational Research Association (AERA), 8, 84
American Educational Research Journal, 6
analogue research, 172
Angrosino, M. V., 69–70
Anthropological Association, 83
anthropological research strategies, 6
Anthropology Field Methods in the Study of Education, 84
applicability, 61, 98, 105
attribute variable, 48
audit trail, 60, 98, 105

Barbules, N. C., 25–26
Becker, H. S., 68–69
behavioral consultation study, 106–8, 160–75; collaborative vs. expert consultant relationship, 162–63, 167, 170; discussion, 170–73; independent variables, 167–69; labeling bias, 161–62; participants, 163–64; procedures, 164–67; results, 169
behavioral science, 1, 4
Berliner, David, 58
between-methods triangulation, 88
blind raters, 166
Blumer, H., 2
Boostrom, R., 69
Borg, W. R., 48

bracketing, 86
Brewer, J., 63

Campbell, D. T., 35–36, 40–41, 49–50, 179n. 1; symbols, 43–44, 107
case-study methods, 74–75
causal comparative design, 48
causality, 12, 43; ex post facto research and, 48–49; historical methods and, 79, 82
chain of reasoning, 11
Charmaz, K., 21, 70–71, 72, 73
coding, 71–72
communicative validity, 55, 56
comparability, 85
complementary methods, 63–64
Comte, Auguste, 3
conclusion-oriented studies, 8
conclusions, critique of, 102, 107–8, 128
concurrent validity, 45–46
consistency, 1, 19, 59, 66, 77; epistemological, 112; persistent observation, 58, 97, 104. *See also* reliability
consistency-questions model, 92–97, *93*, *95–96*, 110–12
constructed effects, 89, 90
constructs, 60
construct validity, 46, 55, 101
consumer of research, 111, 160
context specificity, 49
correlational research, 48. *See also* ex post facto research
Council on Anthropology and Education, 84
credibility, 40, 45, 56, 62, 85
Creswell, J. W., 83

203

criteria, dilemma of, 52
Cronbach, L., 8
Culbert, S. A., 138
Culbertson, J. A., 6
Culture of Schools Program, 83
Curtis, John M., critique of study by, 100–103, 130–42

Daniel, L. G., 52
data, 7, 179n. 4
data-analysis technique, 21, 134–35; grounded theory, 70–74
data collection: reliability in, 46–47; theoretical sampling, 60, 73, 98, 105
decision-oriented studies, 8
deductive reasoning, 22, 23
definitional validity, 45
Denzin, N. K., 24–25, 35, 37, 52, 180n. 4; on qualitative research as term, 19–20
dependent variables, 48, 106, 133–34
derivation of hypotheses, 33
design validity, 36, 40–42; criteria, 57–62
Dewey, John, 4
Diamond, Stanley, 83
Diesing, P., 16
disconfirmability, 4
Donmoyer, R., 61

education: demographic changes in, 143–45; nature of science and research in, 12–15; quantitative paradigm and, 4, 5–6. *See also* monocultural graduate in multicultural environment
Educational Researcher, 8
Egan, K., 112
Eisner, E. W., 37
embodied validity, 54
emerging theory, 73–74
empirical research, 5, 7, 16
epistemological assumptions, 65–66
epistemological consistency, 112
equivalent forms, 42–43, 47

Erickson, Frederick, 30
Ethnographic Interview, The (Spradley), 77
ethnographic methods, 30–33, 83–85
evidence, 10–11, 18, 28
experimental mortality, 47
expert-judge content validity, 45
explanation, 3, 11–12, 21
exploratory research, 48
ex post facto research, 36, 43, 48–50
external criticism, 80–81
external validity, 24, 36, 41–45, 43; threats to, 49–50, 101

face validity, 45
factor analysis, 46
falsifiability, 16, 20–21
feedback loops, 29, 33–34
Fink, A. S., 78
first impressions, 138, 139
focused coding, 71–72, 73
Foerstner, S. B., 87, 119–22
Fontana, A., 14, 77–78
Fraas, J., 23
Frey, J. H., 14, 77–78
Fuller, Mary Lou, critique of study by, 103–6, 143–59

Gall, J. P., 48
Gall, M. D., 48
Gay, L. R., 68
Gearing, Fred, 84
Geer, B., 68–69
Geertz, Clifford, 2
generalizability, 61–62, 85, 172; consistency-questions model and, 92, 94, 98, 105; threats to, 49–50
Gestalt theory, applied to suicide incident, 97–100, 123–29
Glaser, B. G., 21, 70
Goetz, J. P., 42, 84–85, 89–90
Good, C. V., 79–80, 81
Grant, C. A., 144
Guba, E. G., 25, 40, 52, 56; design validity criteria of, 35–36, 57–62

INDEX

Haberman, M., 143
Hakim, C., 76
Hammersley, M., 40, 52–53
Harman, Robert L., critique of study, 97–100, 123–29
historical methods, 5, 44, 78–83
history effects, 89, 90
holistic research, 16, 18–19, 27, 33, 91, 94, 112, *176–77*
Howard, D. C. P., 86–87
Howe, K. R., 26
Hunter, A., 63
Husserl, Edmund, 87
hypotheses, 3, 32–33, 100; case-study methods and, 75; grounded-theory methods and, 70–74; qualitative research and, 20–21; soft hypothesis testing, 60
hypothesis-testing research, 23–24
hypothetico-deductive system, 4

idealism, 5, 179n. 3
independent variables, 48, 106, 133
inductive reasoning, 22, 23, 78
initial codes, 71–72
instrumentalism, 53
instrumentation, 36, 41, 45–47
interactive continuum, 1–2, 9–10, 16–17, *31*, 90; applying, 91–108; benefits of, 29–30; as category of mixed methods research, 27–34; conceptualization of, 17, 22, 23, 29–30; critiquing research, procedures to use, 92–97, *93*, *95–96*. *See also* consistency-questions model; mixed methods research; qualitative-quantitative research
internal criticism, 80–81
internal validity, 24, 36, 41–45; equivalent forms, 42–43; threats to, 36, 44–45, 49, 101
interpretation, 54, 72–73, 77
Intervention Rating Profile-15 (IRF-15), 168–69, 172
interviewing methods, 14, 47, 75–78
investigation of essences, 86
investigation validity, 55–56
involvement of consultee, 106–8, 160–75
ironic validity, 54
Irvine, J. J., 157
iterative process, 27–28, 39, 111–12

Jaffe, P. E., 138
Jick, T. D., 88–89
Johnson, B., 63–64
judgment, 86, 94

Kitzinger, C., 53–54
knowledge, 3–4, 10–11, 28; conflicting ways of knowing, 62–65
known-group validity, 46
Koenig, D., 87, 119–22
Kratchowill, Thomas R., 106–8, 160–75
Krathwohl, D., 11
Kuhn, Thomas S., 5
Kvale, S., 55–57, 78, 86, 87

labeling bias, 161–62
Lagemann, E. C., 8
Lather, P., 54
laughter and humor, study on, 87, 119–22
learning, 4
LeCompte, M. D., 42, 84–85, 89–90
legitimation crisis, 2–3, 24–25, 37, 52
Lieberman, A., 19
Lincoln, Y. S., 24–25, 35–36, 40, 52, 56, 180n. 4; design validity criteria, 35–36, 57–62; phenomenological research, view of, 85–86; on qualitative research as term, 19–20
literature review, 32
lived experience, 87
Lofland, J., 68
logical validity, 45
Lombard, G. J., 90

mathematics, symbolic logic of, 4
maturation, 44

McNeil, K., 23, 76
meaning, 36–37, 51, 73–74
meaning units, 119, 121–22
measurement: behavioral consultation study, 168–69; therapist self-disclosure study, 133–35
measurement validity, 36, 40–41; instrumentation, 45–47
Medawar, P. B., 10
member checking, 59, 98, 105
memo writing, 73
Merriam, S. B., 74
methodism, 52–53
methods, 5, 7, 39; behavioral consultation study, 163–69; links between research questions and truth value, 37–40, *38*; monocultural graduate students study, 145–46; therapist self-disclosure study, 132–35. *See also specific methods*
methods validity, 42
Miller, L., 19
Mitchell, J. G., 87
mixed methods research, 1–2; categories of, 17, 26–34; conceptualization of, 24–26, 29–30; dangers of, 3, 9, 14, 66, 109–10; principles of practice in, 110–14; validity in, 62–65. *See also interactive continuum*
monocultural graduate in multicultural environment, 103–6, 143–59; educational implications, 156–58; themes, 147–56
Morse, J. M., 110
Mouly, G. J., 28, 68; historical methods, view of, 77, 79, 80, 81
multimethod-multitrait research strategies, 88
multiple realities, 2, 53
multiple sites, 89–90
multivariate analyses, 50–52

Naturalistic Inquiry (Lincoln and Guba), 57
naturalists, 3, 6–7, 16–17

negative-case analysis, 62, 98, 106
neutrality, 58, 97, 103
Newman, I., 23, 41–42, 76, 87, 119–22
New School for Social Research, 83
new synthesis, 19
No Child Left Behind Act (NCLB), 8–9
nomological network, 53, 60, 75
nonhuman phenomena, 51
nonintegrated research, 26, 27
Norell, Mark, 10

objectivity, 24, 69
observational methods, 67–70
Ödman, P. J., 82–83
Onwuegbuzie, A. J., 52
open description, 86
organizational theory, 4
outcome measure, 49

Page, R. N., 53
paradigms, 5, 179n. 6
paralogical validity, 54
partially structured interviews, 76
participant observation, 67–69, 84
participants: behavioral consultation study, 163–64; in graduate teacher study, 146–47; groups in suicide study, 124–25; meaning of phenomena for, 74; monocultural graduate students study, 146–47; small-group sessions, 126–27; therapist self-disclosure study, 132
participant-to-variable ratio, 51
Patton, M. Q., 20, 21, 51, 56, 74; interviewing methods, 75–76
peer debriefing, 58–59, 77, 97, 104
persistent (consistent) observation, 58, 97, 104
Peshkin, A., 37
phenomenological reduction, 86
phenomenological research, 2, 3, 85–87; enhancing validity of, 87–90; laughter and humor study, 119–22
Phillips, D. C., 25–26

INDEX

Piaget, Jean, 5, 61
policymaking, 6
Polkinghorne, D. E., 40, 55, 57, 61
Popper, Karl, 16
positivism, 3–4, 19; logical, 4–6; methodism, 52–53. *See also* truth
postmodernism, 24, 53–55
postpositivism, 17, 24–26, 109, 113–14, 180n. 4; validity and, 39–40
poststructuralists, critical, 24
poverty, 143–44, 150–51
pragmatism, 8, 26, 56, 63
prediction, 78
predictive validity, 46
pretest, as factor in validity, 44–45, 47, 50
principles of practice, 109, 110–14
process: grounded theory, 71; iterative, 27–28, 39, 111–12
Program in Anthropology and Education (PAE), 84
prolonged engagement on-site, 58, 97, 103–4
proportionate reason, 69–70
published research, critique of, 91–97; Alexander and Harman study, 97–100, 123–29; Curtis study, 100–103; Fuller study, 103–6; reflections, 99–100, 103, 106, 108; Rhoades and Kratchowill study, 106–8

Qualitative Inquiry in Education (Eisner and Peshkin), 36–37
qualitative-quantitative continuum. *See* interactive continuum; mixed methods research
qualitative-quantitative research, as false dichotomy, 2, 9–10
qualitative research: ascendance of, 6–7; case-study methods, 74–75; conceptualization of, 19–23, 29; critique of published research, 97–100; design validity criteria, 35–36, 57–62; ethnographic methods, 30–33, 83–85; generic definition, 19–20;

grounded-theory methods, 21, 70–74; historical methods, 78–83; interviewing methods, 75–78; observational methods, 67–70; phenomenological research, 85–90; philosophies of, 52–53; recommendations, 102–3, 108; theory and, 20–21; truth and, 52–54, 62; validity in, 52–57
quality of craftsmanship, 55–56
quantitative research: conceptualization of, 23–24, 29; critique of published research, 100–103; generalizability, 49–50, 61–62; historical methods and, 81–82; as hypothesis-testing research, 23–24; methods, 7; validity in, 40–52. *See also* positivism

randomized trial studies, 7, 9
reactive observation, 67–68, 70
reader understanding, 74
reality, 2–4, 53
reciprocity, 138–39
referential materials, 59, 98, 105
relational communication analysis, 163
relationships, ex post facto research and, 49
relativism, 53, 81
reliability, 24, 40, 53, 87; in data collection, 46–47. *See also* consistency
replicability, 11, 23–24, 46–47; critique of, 60, 98, 105; generalizability and, 61–62; validity and, 41–42
research design, 36, 39, 67, 91, 101; behavioral consultation study, 168; decisions about, 17–18, 110–11; therapist self-disclosure study, 132–33; univariate and multivariate analyses, 50–52; validity criteria for, 35–36, 57–62
researcher, 21, 25, 71; as instrument of data collection, 21; skills of, 72–73, 76, 77; validity of mixed methods research and, 62–65

researcher bias, 30–32, 68, 72, 77
research in education, 12–15
research paradigms, evolution of, 2–9
research purpose, 1, 17, 90; interview types and, 77; iterative process, 27–28, 39, 111–12; links between methods and truth value, 37–40, 38; principles of practice and, 110; systematic approach, 17–19; typology of, 30–32; validity and, 35
research question, 1, 90; case-study methods and, 74; critique of, 100, 101, 106; iterative process, 27–28, 39, 111–12; links between methods and truth value, 37–40, 38; principles of practice and, 110; role of, in design, 17; in social sciences, 7–8
results: behavioral consultation study, 169; laughter and humor study, 120–21; therapist self-disclosure study, 135, 135–37, 136
rhizomatic validity, 54
Rhoades, Mary M., 106–8, 160–75
Rhoades and Kratchowill study, critique of, 106–8
rigor, 56, 57

sampling, 43; theoretical, 60, 73, 98, 105
Sandelowski, M., 64–65
Schatzman, L., 77
science: in education, 12–15; as foundation for research design, 10–12; historical methods as, 79–80, 81; as holistic, 16, 18–19; as set of systematic procedures, 17–19; as way of knowledge, 10–11
selection bias, 47, 50
selection effects, 89
self-correction, 10, 11, 26, 29
self-selection, 49
sensory input, 3
setting effects, 89–90
settings, 49
Shadish, W. R., 49

Shafer, R. J., 79, 80, 81, 82
Shulman, Lee S., 11
simultaneous attempt, 26, 27
Sleeter, C. E., 144
Smith, L. M., 30–32
social sciences: positivist, 3–4; research questions in, 7–8
Sources of a Science of Education (Dewey), 4
specifications, table of, 45
speculation, 30
Spencer, Herbert, 3
Spindler, G. D., 84
Spradley, J. P., 77
Srinivasan, R., 90
Stake, R. E., 74–75
standards of practice, 110, 113. See also principles of practice
Stanford Conference (1954), 6
Stanley, J. C., 35–36, 40–41, 49–50, 179n. 1; symbols, 43–44, 107
statistical analysis, 7, 79, 102
statistical (empirical, criterion) validity, 46
statistical generalizability, 61
statistical regression, 47
statistical significance, 41–42
Stivers, E., 90
Strauss, A. L., 21, 70, 77
structural relationships, 59–60, 98, 105
structured interviews, 47, 76
Structure of Scientific Revolutions, The (Kuhn), 5
suicide study, 97–100, 123–29
symbols, research, 43–44, 107
synthesis, 81
systematic approach, 17–19
systematic knowledge, 10–11

Tashakkori, A., 8, 25, 53, 56
teacher education programs, 144–45; advice for preservice teachers, 154–55
Teddlie, C., 8, 25, 56, 63

testing factors, 44–45
themes: laughter and humor study, 121–22; monocultural graduate in multicultural environment, 147–56
theoretical sampling, 60, 73, 98, 105
theory, 3, 20–21, 29; revision, 33, 34
therapist self-disclosure study, 100–103, 130–42; discussion, 137–39; methods, 132–35; research design, 132–33; results, *135*, 135–37, *136*
thick description, 92
Tierney, W. G., 53–54
Toma, J. D., 56
topic of research, 32
traditional, as term, 12
transcripts, 78
transferability, 61–62, 98, 105
translatability, 85
treatment, forms of, 49–50
treatment effects, 42–43
triangulation, 56, 59, 75, 88–89; critique of, 97, 104–5
trustworthiness, 24, 37
truth, 3, 6–7, 10; correspondence theory of, 54–55; qualitative research and, 52–54, 62, 98–99
truth value, 37; critique of, 98–99, 106; links between research questions and methods, 37–40, *38*; measurement validity and, 46; rigor and, 57
Turner, L. A., 53–64

understanding, 36, 57
unit of measurement, 49
univariate analyses, 50–52
unobtrusive observation, 67–68, 70
unstructured interviews, 76
U.S. Office of Education (USOE), 83

validity, 1, 24, 30; of case-study methods, 75; construct validity, 46, 55, 101; critique of, 101, 107; design validity criteria, 35–36, 57–62; epistemological assumptions, 65–66; in ethnographic methods, 84–85; evolution of emphasis on, 36–37; external and internal criticism, 80–81; of grounded-theory methods, 71–74; of historical methods, 82–83; instrumentation and, 45–47; internal and external, 24, 36, 41–45, 49–50, 101; of interviewing methods, 77–78; meaning of, 36–37; methods validity, 42; in mixed methods research, 62–65; multiple sites and, 89–90; of observational methods, 69–70; postpositivism and, 39–40; pretest as factor in, 44–45, 50; in qualitative research, 52–57; in quantitative research, 23, 40–52; replicability and, 41–42; as social construct, 53–54; thirteen criteria for, 35–36, 40; threats to, 36, 44–45, 49–50, 82, 101; truth value and, 37, *38*; types, 54–56
validity check, 80
values, 4, 6–7, 25
Van Manen, M., 2, 86
variables, 48, 50–52, 106–7, 133–34
variance design, 132–33
verifiability, 11, 16
videotapes, 165–66
Vienna Circle, 4
Villegas, A. M., 157

Wax, M. L., 84
Wertz, E. J., 87
Winter, G., 35, 57
within-methods triangulation, 88
Wittgenstein, Ludwig, 112
Wolcott, H. F., 36–37
working hypotheses, 32

Zeichner, K. M., 157

Carolyn S. Ridenour is a professor in the Department of Educational Leadership at the University of Dayton, where she regularly teaches qualitative and quantitative research methodology courses and directs doctoral dissertations. In addition to mixed methods research, she has published in the areas of urban schools, gender and education, and cultural diversity. (Before 1999, she published as Carolyn R. Benz.)

Isadore Newman is an emeritus distinguished professor at the University of Akron, where he taught advanced research courses and mixed methodology. He currently is visiting director of graduate studies and research at Florida International University. He is the author of *Testing Research Hypotheses with the General Linear Model* and *How to Be Involved in Program Evaluation: What Every Administrator Needs to Know.*